Mycorrhizas:

Anatomy and Cell Biology

R. Larry Peterson
Department of Botany
University of Guelph
Guelph, ON N1G 2W1

Hugues B. Massicotte
University of Northern British Columbia
Ecosystem Science and Management Program
3333 University Way
Prince George, BC V2N 4Z9

Lewis H. Melville
Department of Botany
University of Guelph
Guelph, ON N1G 2W1

NRC-CNRC
NRC Research Press
Ottawa 2004

CABI *Publishing*

Printed in Canada on acid-free paper.

Published exclusively in Canada and non-exclusively in other territories of the world (excluding Europe, Asia, Africa, the Middle East, Australia & Oceania and South America) by NRC Research Press as ISBN 0–660–19087–7.
NRC No. 46325

Published exclusively in Europe, Asia, Africa, the Middle East, Australia & Oceania and South America, and non-exclusively in other territories of the world (excluding Canada) by CABI Publishing, a division of CAB International, with the ISBN 0–85199–901–8.

CABI Publishing, CAB International, Wallingford, Oxon OX10 8DE UK

Tel: +44 (0)1491 832111 Fax: +44 (0)1491 833508
E-mail: cabi@cabi.org Web site: www.cabi-publishing.org

National Library of Canada cataloguing in publication data

Peterson, R. Larry

Mycorrhizas : anatomy and cell biology

Includes bibliographical refere.ıces and an index.
Issued by the National Research Council of Canada.
NRC No. 46325
ISBN 0–660–19087–7

1. Mycorrhizas. 2. Roots (Botany)
I. Massicotte, Hugues B. II. Melville, Lewis H. III. National Research Council Canada. IV. Title.

QK604.2.M92P37 2003 581.4'98 C2003-980254-X

NRC Monograph Publishing Program
Editor: P.B. Cavers (University of Western Ontario)

Inquiries: Monograph Publishing Program, NRC Research Press, National Research Council of Canada, Ottawa, Ontario K1A 0R6, Canada. Web site: www.monographs.nrc-cnrc.gc.ca

Contents

Preface

The occurrence of symbiotic relationships between roots and fungi (mycorrhizas) has been recognized since the early 1800's and it is now clear that these associations are the most prevalent symbiotic systems on earth. Mycorrhizal associations can be found in all ecosystems and in important forest and crop species. Based on the structural features of the symbiosis between the plant and the fungus involved, seven categories have been described. Ongoing and more intensive investigations of mycorrhizas in the field and in regions of the world that have received little attention in the past, will likely result in the present categories being modified and perhaps expanded.

This book has resulted from a long collaboration among the authors and our shared interest in exploring the structural diversity of mycorrhizas at various scales. It has become clear from the literature and recent international meetings that a majority of the current research on mycorrhizas involves molecular, physiological and ecological studies. Few structural studies are being pursued in spite of the need for basic information for some categories of mycorrhizas and the detailed information that can now be obtained as a result of new analytical methods. The intention of this book, therefore, is to provide, through a series of chapters, a summary of all the mycorrhizal categories from a morphological and anatomical perspective. We strongly believe that the structural information presented will provide researchers in all aspects of mycorrhizal research with a platform from which to explore questions related to the functioning of mycorrhizas.

This book will be of interest to undergraduate and graduate students, secondary school teachers and professors who include discussions of mycorrhizas in various subjects that they teach, and researchers in root biology and various aspects of mycorrhizas. We hope that the information included and the format of the book will stimulate readers to explore mycorrhizas in more detail. Leads are given into the literature for further discussion of each topic covered.

Most of the images presented in the chapters are from our own research. Credit is given for those images provided by other researchers and we thank them for allowing us to use them. Although some references are included in each chapter, no attempt has been made to provide a comprehensive literature review for each topic. Several recent reviews and volumes on mycorrhizas provide excellent detailed discussions of various aspects of mycorrhizal associations.

To provide the most complete descriptions of mycorrhizas, information has been gained by using a variety of preparative techniques combined with a range of microscopical methods. A brief appendix describing the various techniques used in preparing the images in the book is included, primarily as a guide for readers who are new to the field or have a background that does not include structural training. Also included is a glossary of the most common terms related to structural aspects of mycorrhizas.

Financial support has been provided indirectly by the Natural Sciences and Engineering Research Council of Canada by research grants over many years. We thank all of our former and present undergraduate, graduate, and post-doctoral students for their excellent contributions, Cameron Ackerley and Melanie Howarth who, for years, helped us with many aspects of microscopy, Linda Tackaberry for editorial assistance, Christopher Peterson for helping organize our literature data base, and the Universities of Guelph and Northern British Columbia for providing RLP and HBM time to launch this project.

Introduction

Root–microorganism interactions are ubiquitous

Roots of all vascular plants are in intimate contact with a substrate, and whether this is soil, water or tree bark, various abiotic and biotic factors influence their development and function (Peterson 1992). A complex array of microorganisms occupies various niches in this substrate and this affects roots and, therefore, plant performance in various ways. For example, many of these organisms are pathogenic fungi or bacteria that are detrimental to plant growth and reproduction. Consequently, considerable effort and expense may be required to control the population density and (or) pathogenicity of these organisms, particularly if they attack crop, horticultural, or forest species. Other organisms may be beneficial to plants and these may be manipulated in ways to increase their effect on plant performance. Examples include plant growth promoting rhizobacteria (PGPRs), a group of free-living bacteria that may fix nitrogen, provide growth-promoting substances to plants, or modify the rhizosphere to make various recalcitrant nutrients available to plants. Various members of the Rhizobiaceae (e.g., *Rhizobium*, *Bradyrhizobium*) and the actinomycete genus *Frankia* are capable of inducing nodules on legumes and several woody species, respectively, and once housed within nodules, are able to fix atmospheric nitrogen that can be made available to the plant.

The most prevalent beneficial organisms associated with plants, however, are soil-inhabiting fungi that form mutualistic root–fungal associations referred to as mycorrhizas (Smith and Read 1997). There has been a long history in studying mycorrhizas (Trappe and Berch 1985; Harley and Smith 1983) and this early work formed a sound basis for the proliferation of research over the past few decades and for the number of books published that are devoted to various aspects of mycorrhizas. Of the recent books, Smith and Read (1997) is the most complete in terms of coverage of all categories of mycorrhizal associations; these authors consider structure, physiology and some ecological aspects. Other books are more specialized in their coverage and are listed at the end of this section for reference.

Mycorrhizal categories

Since this volume is concerned with the structure of all known mycorrhizal categories, a brief definition of each is given here. The pioneering work of Frank (see Smith and Read 1997) resulted in the recognition of two broad subdivisions of mycorrhizas, ectomycorrhizas and endomycorrhizas. Ectomycorrhizas are characterized by the formation of a mantle and a Hartig net of intercellular hyphae on roots of predominantly tree species. Endomycorrhizas are more variable than ectomycorrhizas in that herbaceous and tree species are involved and there is a diversity of fungal groups involved in forming these associations. Endomycorrhizas have, therefore, been further classified as either arbuscular mycorrhizas, ericoid mycorrhizas, arbutoid mycorrhizas, monotropoid mycorrhizas, ectendomycorrhizas or orchid mycorrhizas. Each of these categories is characterized by the invasion of root cells by fungal hyphae but profound differences occur in the nature of intracellular hypha development.

Arbuscular mycorrhizas (formerly referred to as vesicular–arbuscular mycorrhizas) are by far the most prevalent of all mycorrhizal categories with more than 80% of all plant species showing an association involving a few fungal genera in the Glomeromycota. The most diagnostic feature of this mycorrhiza is the development of intercellular hyphae, intracellular hyphae and arbuscules in root cortical cells (some species also develop vesicles within and between root cells), and the production of spores on intraradical and extraradical hyphae.

Ericoid mycorrhizas are found in the families Ericaceae and Epacridaceae of the angiosperm order, Ericales. This type is extremely important in heathland ecosystems where soil nitrogen is bound in various organic compounds. The nitrogen in these compounds is accessed by the plant primarily via the fungi associated with fine roots. Epidermal cells of these "hair roots" are colonized by fungal hyphae which form intracellular hyphal complexes. The few identified fungal species belong primarily to the Ascomycotina.

Arbutoid and monotropoid mycorrhizas are also found in members of the Ericales; these specialized mycorrhizas differ structurally from ericoid mycorrhizas by having a Hartig net as well as intracellular hyphae and by involving a different suite of fungal partners. Monotropoid mycorrhizas are distinct from arbutoid mycorrhizas in that epidermal cells are invaded by a single hypha forming a 'peg' around which the host cell elaborates a wall and plasma membrane. Arbutoid mycorrhizas, on the other hand, develop a hyphal complex in epidermal cells.

Ectendomycorrhizas resemble ectomycorrhizas, arbutoid, and monotropoid mycorrhizas by having a mantle and Hartig net, however, they are confined to the conifer genera *Pinus* and *Larix* and are formed by a small group of Ascomycete fungi (Yu et al. 2001).

Orchid mycorrhizas are restricted to the large angiosperm family, Orchidaceae, and are unique in that fungal associations occur with embryo cells of germinating seeds (Peterson et al. 1998) as well as with roots of seedlings and mature plants. In both situations, various fungal species belonging to the Basidiomycota, form intracellular coils called "pelotons".

Dark septate fungal endophytes

We have included a chapter on the association of dark septate fungal endophytes with plant roots because of their prevalence in various ecosystems, and because it has been demonstrated that some may form mutualistic relationships with certain plant species.

Microscopy

Some of the many advances in studying the structural changes that occur in both plant and fungal symbionts during the formation of mycorrhizal associations have been made possible by improvements in microscopes, the development of new preparative methods for scanning and transmission electron microscopy, the development of confocal laser scanning microscopy, and the use of specialized methods for the localization of specific compounds at the cellular and subcellular level. We have provided a brief introduction to some of these methods in the appendices and provide references to textbooks for detailed preparative methods.

Organization of book

This book is organized into eight chapters, the first seven dealing with the morphology and anatomy of the categories of mycorrhizas and chapter eight dealing with dark septate fungal endophytes of roots. The frontispiece for each chapter is intended to provide a sense of the plant landscapes involved in each mycorrhiza category. Specialized topics, most of which relate to microscopy in advancing our understanding of the functioning of mycorrhizas, are included within "boxes" in some chapters. A brief appendix of methods, a glossary of terms, and an index are also included.

References

Harley, J.L., and Smith, S.E. 1983. Mycorrhizal Symbiosis. Academic Press, London, U.K.

Peterson, R.L. 1992. Adaptations of root structure in relation to biotic and abiotic factors. Can. J. Bot. **70:** 661–675.

Peterson, R.L., Uetake, Y., and Zelmer, C. 1998.

Fungal symbioses with orchid protocorms. Symbiosis, **25:** 29–55.

Smith, S.E., and Read, D.J. 1997. Mycorrhizal Symbiosis. Second Edition. Academic Press, San Diego, Calif., U.S.A.

Trappe, J.M., and Berch, S.M. 1985. The prehistory of mycorrhizae: A.B. Frank's predecessors. *In* Proceedings of the 6th North American Conference on Mycorrhizae. *Edited by* R. Molina. Forest Research Laboratory, Corvallis, Oreg., U.S.A.

Yu, T.E.J.-C., Egger, K. N., and Peterson, R.L. 2001. Ectendomycorrhizal associations – characteristics and functions. Mycorrhiza, **11:** 167–177.

Recent specialized books on mycorrhizas

Allen, M.F. (*Editor*). 1992. Mycorrhizal Functioning, An Integrative Plant Fungal Process. Chapman & Hall, New York, U.S.A.

Brundrett, M., Bougher, N., Dell, B., Grove, T., and Malajczuk, N. 1996. Working with Mycorrhizas in Forestry and Agriculture. The Australian Centre for International Agricultural Research (ACIAR) Monograph 32. Canberra, Australia.

Brundrett, M., Melville, L., and Peterson, L. (*Editors*). 1994. Practical Methods in Mycorrhiza Research. Mycologue Publications, Waterloo, Ont., Canada.

Cairney, J.W.G., and Chambers, S.M. (*Editors*). 1999. Ectomycorrhizal Fungi, Key Genera in Profile. Springer-Verlag, Berlin, Germany.

Dighton, J. 2003. Fungi in Ecosystem Processes. Marcel Dekker Inc., New York.

Fortin, J.A., Charest, C., and Piché, Y. 1995. La symbiose mycorhizienne, état des connaissances. Orbis Publishing.

Gianinazzi, S., Schüepp, H., Haselwandter, K., and Barea, J.M. 2002. Mycorrhizal Technology in Agriculture. From Genes to Bioproducts. Birkhaüser Verlag AG, Heidelberg, Germany.

Kapulnik, Y., and Douds, D.D. Jr. (*Editors*). 2000. Arbuscular Mycorrhizas: Physiology and Function. Kluwer Academic Publishers, Dordrecht, The Netherlands.

Norris, J.R., Read, D.J., and Varma, A.K. (*Editors*). 1991. Methods in Microbiology, Vol. 23. Techniques for the Study of Mycorrhiza. Academic Press, London, U.K.

Pfleger, F.L., and Linderman, R.G. (*Editors*). 1994. Mycorrhizae and Plant Health. APS Press, St. Paul, Minn., U.S.A.

Podila, G.K., and Douds, D.D. Jr. (*Editors*). 2000. Current Advances in Mycorrhizae Research. APS Press, St. Paul, Minn., USA.

Rasmussen, H.N. 1995. Terrestrial Orchids from Seed to Mycotrophic Plant. Cambridge University Press, Cambridge, U.K.

Read, D.J., Lewis, D.H., Fitter, A.H., and Alexander, I.J. (*Editors*). 1992. Mycorrhizas in Ecosystems. C.A.B. International, Wallingford, U.K.

Robson, A.D., Abbott, L.K., and Malajczuk, N. (*Editors*). 1994. Management of Mycorrhizas in Agriculture, Horticulture and Forestry. Kluwer Academic Publishers, Dordrecht, The Netherlands.

Sharma, A.K., and Johri, B.N. (*Editors*). 2002. Arbuscular Mycorrhizae. Interactions in Plants, Rhizosphere and Soils. Science Publishers Inc., Enfield, N.H., U.S.A.

Smith, S.E., and Read, D.J. 1997. Mycorrhizal Symbiosis. Second Edition. Academic Press, San Diego, Calif., U.S.A.

Smith, S.E., and Smith, F.A. (*Editors*). 2002. Diversity and Integration in Mycorrhizas. Kluwer Academic Publishers, Dordrecht, The Netherlands.

van der Heijden, M.G.A., and Sanders, I. R. (*Editors*). 2002. Mycorrhizal Ecology. Ecological Studies, Vol. 157. Springer-Verlag, Berlin, Germany.

Varma, A. (*Editor*). 1998. Mycorrhiza Manual. Springer-Verlag, Berlin, Germany.

Varma, A., and Hock, B. (*Editors*). 1998. Mycorrhiza Structure, Function, Molecular Biology and Biotechnology. Second Edition. Springer-Verlag, Berlin, Germany.

Mycorrhizas – the most prevalent symbioses on earth

Most of the research on the interactions between plants, the environment, and other organisms has involved those parts of plants that are visible and easy to access. However, the underground system, mainly roots, interacts with the soil environment and with a multitude of organisms in very complex ways. The interaction of roots with mycorrhizal fungi is almost ubiquitous in both natural and man-made ecosystems. Because of this, there are a number of reasons why the study of mycorrhizas is important:

- Mycorrhizas increase nutrient uptake from the soil.
- Mycorrhizal fungi can be used in the biocontrol of pathogenic fungi and nematodes.
- Some mycorrhizal fungi can bind heavy metals thus protecting plants from toxic levels of these substances.
- Mycorrhizas are useful in the restoration of degraded sites.
- Mycorrhizas have a positive effect on the establishment of plant communities.
- In orchids, and other myco-heterotrophic species, mycorrhizal fungi are essential for seed germination and seedling establishment.
- Fruitbodies [e.g., many truffles, chanterelles, *Lactarius* and other species] associated with many mycorrhizal fungi are edible by humans and other organisms.

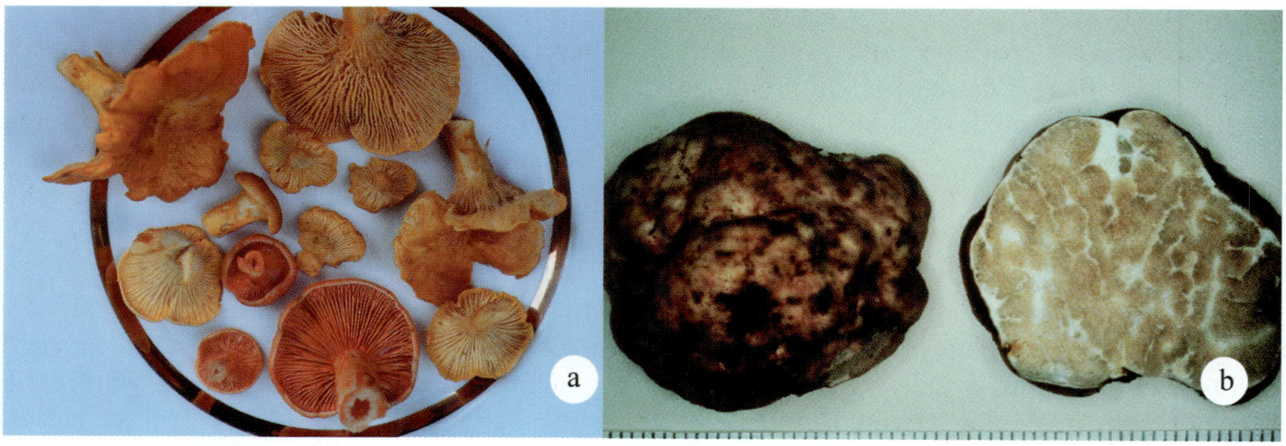

Figure a. Dish of chanterelles (*Cantharellus cibarius*) and *Lactarius deliciosus* ready for consumption. **Figure b.** The succulent Oregon white truffle (*Tuber gibbosum*) sectioned to reveal internal features. (Photo courtesy of J. M. Trappe, 5641).

Chapter 1. Ectomycorrhizas

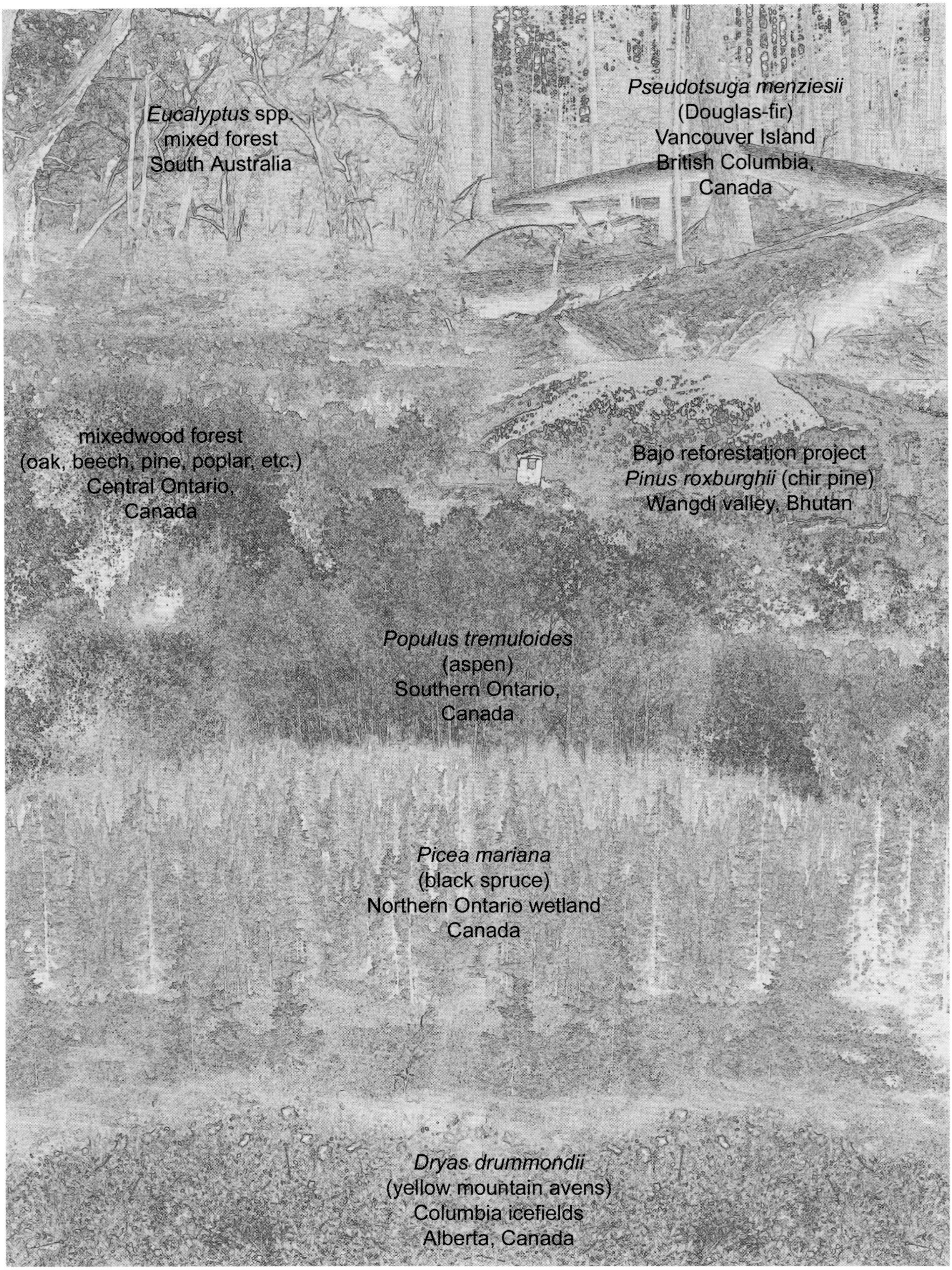
Eucalyptus spp.
mixed forest
South Australia
Pseudotsuga menziesii
(Douglas-fir)
Vancouver Island
British Columbia,
Canada
mixedwood forest
(oak, beech, pine, poplar, etc.)
Central Ontario,
Canada
Bajo reforestation project
Pinus roxburghii (chir pine)
Wangdi valley, Bhutan
Populus tremuloides
(aspen)
Southern Ontario,
Canada
Picea mariana
(black spruce)
Northern Ontario wetland
Canada
Dryas drummondii
(yellow mountain avens)
Columbia icefields
Alberta, Canada

Chapter 1. Ectomycorrhizas

A. Introduction

1. Definition

There is considerable variation in morphological and structural characteristics of ectomycorrhizas, although three features are generally recognized to typify this association: the formation of a mantle or sheath of fungal hyphae that covers considerable portions of lateral roots, the development of hyphae between root cells to form a complex highly branched structure called the Hartig net, and hyphae that emanate from the mantle and grow into the surrounding soil (extraradical mycelium). Figures 1 and 2 illustrate these features in longitudinal and transverse view, respectively. In addition, some ectomycorrhizas develop linear aggregations of hyphae (rhizomorphs or strands), in the extraradical mycelium that are specialized for rapid transport of nutrients and water. A few ectomycorrhizal fungi develop sclerotia consisting of compact storage hyphae surrounded by a rind. Hypogeous or epigeous reproductive bodies may also be formed periodically from extraradical mycelium.

2. Plant species involved

Ectomycorrhizas are usually found on tree species although a few shrub and herbaceous species may also develop this association. Considerable attention has been focused on ectomycorrhizas because many of the tree species involved are important commercially for lumber and paper products worldwide. A detailed list of the conifer and angiosperm families known to have ectomycorrhizas can be found in Smith and Read (1997). Conifer genera such as *Picea* (spruce), *Pinus* (pine) and *Larix* (larch, tamarack) form vast tracts of the boreal forest in the northern hemisphere whereas other conifer species such as *Pseudotsuga menziesii* (Douglas-fir) and *Picea sitchensis* (sitka spruce) occur in the northwestern cool rainforests in North America. Ectomycorrhizal species in angiosperm genera, including *Alnus* (alder), *Betula* (birch), *Fagus* (beech), and *Quercus* (oak), occur widely in north-temperate forests of the world. In the southern hemisphere, the very large angiosperm genus, *Eucalyptus*, and several genera in the family Dipterocarpaceae are the dominant ectomycorrhizal tree species. Due to research efforts in

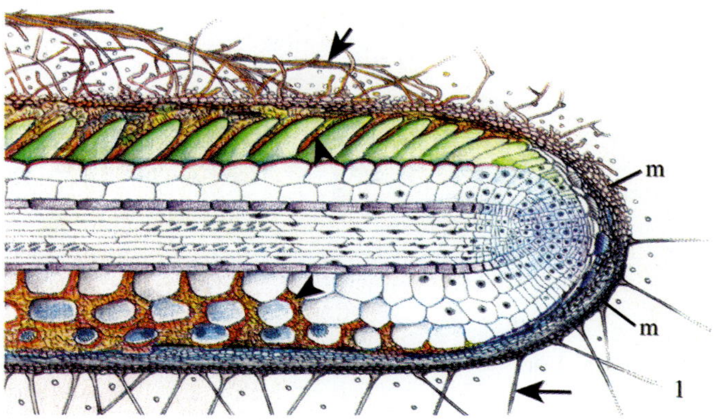

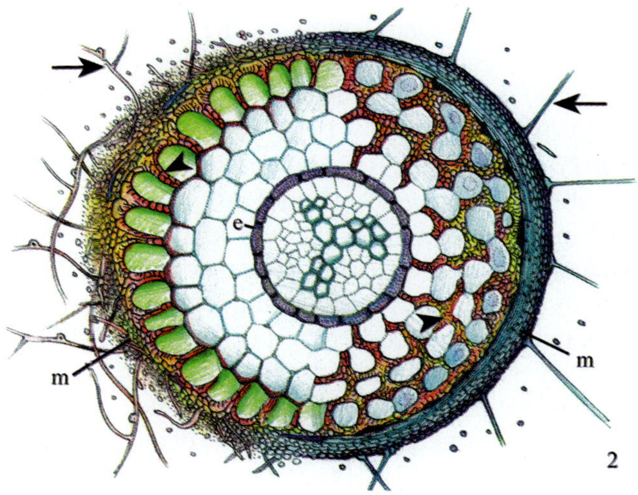

Figure 1. Diagram of ectomycorrhiza in longitudinal section illustrating the major features that occur in angiosperms (top half) and conifers (bottom half). Both have a mantle (m), Hartig net (arrowheads), and extraradical mycelium (arrows). The main difference between these two systems is that the Hartig net in angiosperms is usually confined to the epidermis whereas in conifers it forms around both epidermal and cortical cells.

Figure 2. Diagram of ectomycorrhiza in transverse section illustrating the features of angiosperms (left portion) and conifers (right portion). Extraradical mycelium (arrows), mantle (m), and Hartig net (arrowheads) are indicated. In conifers, Hartig net hyphae are blocked from entering the vascular cylinder by the endodermis (e).

the southern hemisphere, additional ectomycorrhizal species, including many genera of woody legumes, are being described.

Recent fossil evidence of roots of a *Pinus* sp. discovered in the Princeton cherts of British Columbia, Canada, shows that the association between fungi and this conifer genus has existed for over 50 million years (LePage et al. 1997). Dichotomous branching typical of ectomycorrhizal roots of pine species is evident in the fossil material (Figure 3); structural details indicate the presence of hyphae between cortical cells, suggestive of a Hartig net, as well as occasional mantle hyphae (Figures 3, 4).

3. Fungal species involved

The majority of fungal species involved in the ectomycorrhiza symbiosis belong to families in the Basidiomycotina (basidiomycetes), with a few species belonging to the Ascomycotina (ascomycetes). One genus in the Zygomycotina, *Endogone*, can form ectomycorrhizas (Smith and Read 1997). Extensive lists of fungal species can be found in Molina et al. (1992) and Smith and Read (1997). Approximately 5,500 known species of fungi are able to form ectomycorrhizas. Of these, about 80% are epigeous, having reproductive structures occurring above ground (Figures 5–13); fewer species are hypogeous, producing reproductive structures that remain underground (Figures 14–22). The diversity in above ground species is immense and many of the 'mushrooms' found associated with trees are the reproductive structures of ectomycorrhizal fungi (Figures 5–13). Hypogeous species, including the truffles (Figures 14–22), are not as easily detected and more challenging to locate. Both epigeous and hypogeous reproductive structures are often used as indicators of the fungal species present on neighbouring tree roots; however, with the increasing use of molecular techniques, it is now apparent that the diversity of fungi on roots is far greater than the populations of reproductive structures observed.

Reproductive bodies of hypogeous and epigeous fungi are often consumed by small animals (e.g., mammals, marsupials, birds) as sources of food; spores of these fungal species can be passed via faeces forming a part of the life cycle of these organisms. Reproductive bodies are also used as a food source by snails, nematodes and some insects. Some fungal species can associate with a broad range of host species, whereas others are more specialized in the hosts that they can colonize (Molina et al. 1992). The ecological importance of fungi linking many compatible hosts in an ecosystem is receiving considerable attention.

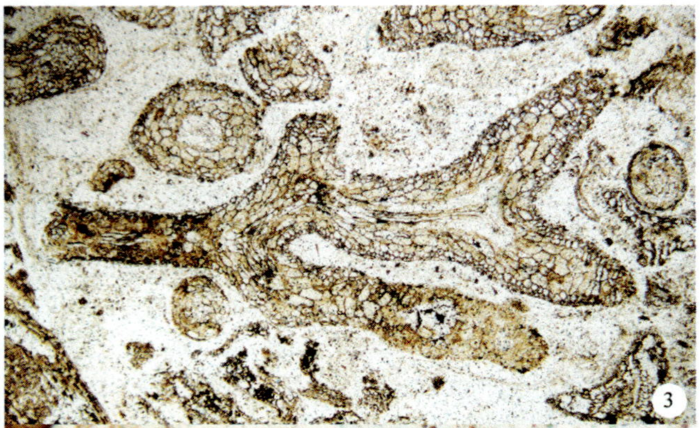

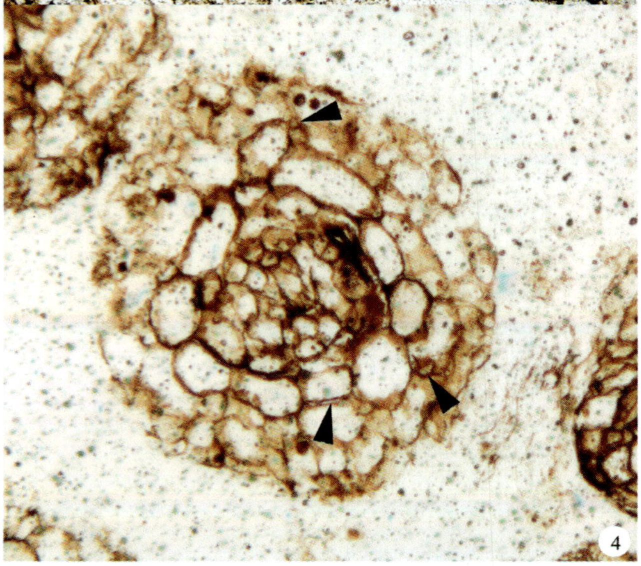

Figure 3. Longitudinal section of fossil *Pinus* ectomycorrhiza from the middle Eocene Princeton chert of southern British Columbia, Canada. A dichotomous root tip is evident.

Figure 4. Transverse section of fossil *Pinus* ectomycorrhiza from the same material as in Figure 3 showing the excellent preservation of root structure and remnants of the Hartig net (arrowheads). Both reprinted with permission from LePage et al. Am. J. Bot. **84**: 410–412 (1997).

B. Morphology of ectomycorrhizas

Ectomycorrhizas can be readily distinguished from uncolonized roots when fine roots are observed in the litter layer and in mineral soil beneath trees. It is evident when examining tree root systems that there is an interaction between host and fungal

23
24
25
26
27
28
29
30
31

Figures 32–38. Mantle features of various ectomycorrhizas.

Figure 32. Unknown morphotype on mature *Pseudotsuga menziesii*. Mantle exhibits long straight cystidia; a basal clamp connection (arrowhead) is evident. (Differential interference contrast (DIC) microscopy).

Figure 33. Unknown morphotype on mature *Pseudotsuga menziesii*. Curling cystidia with thick walls emanate from the mantle surface (DIC microscopy).

Figure 34. Unknown morphotype (possibly a member of the Russulaceae) on mature *Pseudotsuga menziesii*. The mantle surface has numerous ampoule-shaped cystidia, some visible in transverse section. (DIC microscopy).

Figure 35. Unknown dark brown morphotype on *Pseudotsuga menziesii* sapling. The mantle has clusters of raised round cells and emanating hyphae.

Figure 36. Typical stellate pattern of the mantle of *Cenococcum* on *Pinus sylvestris*.

Figure 37. Outer mantle of unknown fungal species (possibly a member of the Russulaceae) on hybrid spruce showing interlocking irregular synenchyma.

Figure 38. Mantle surface showing angular regular synenchyma of a *Tomentella*-like morphotype on hybrid spruce.

Photos in Figures 37 and 38 courtesy of L. Tackaberry.

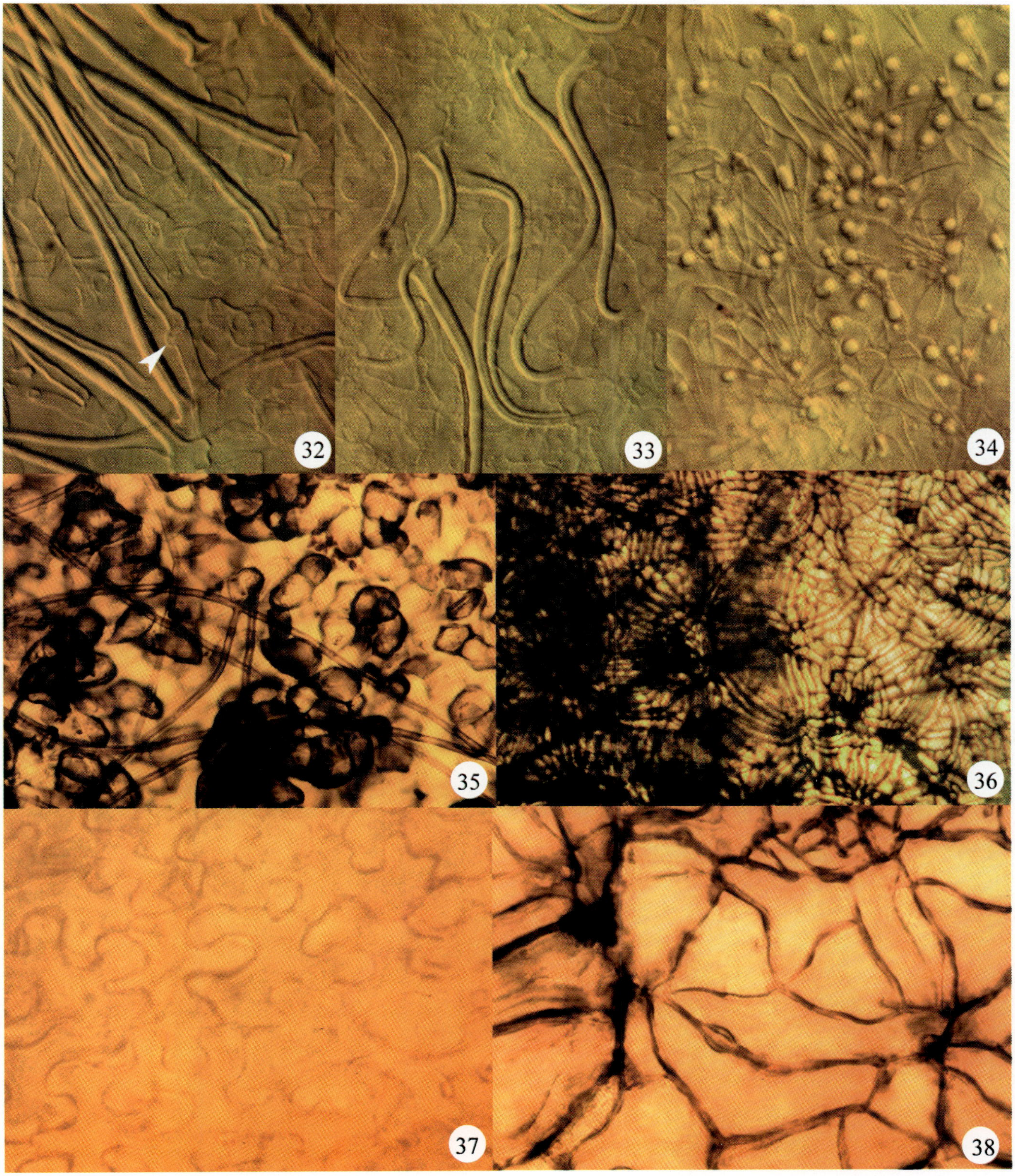
32
33
34
35
36
37
38

Figures 39–43. Sequence in the colonization of root hairs of *Picea mariana* (black spruce) by hyphae of the ectomycorrhizal fungus, *Pisolithus tinctorius.* Black spruce seedlings were grown in growth pouches and root systems inoculated with agar plugs containing mycelium of *P. tinctorius.* Colonized roots were examined by scanning electron microscopy. Figures 41-43 from Thomson et al. Am. J. Bot. **76:** 632–636 (1989).

Figures 44–46. Sectional views of root hairs.

Figure 44. Root hairs (arrowheads) of *Betula alleghaniensis* (yellow birch) colonized by the ectomycorrhizal fungus, *Laccaria bicolor*.

Figures 45, 46. Root hairs of *Pseudotsuga menziesii-Rhizopogon parksii* ectomycorrhizas. The wall of some hairs are thickened (Figure 46).

Figure 46 from Massicotte et al. New Phytol. **147:** 389–400 (2000).

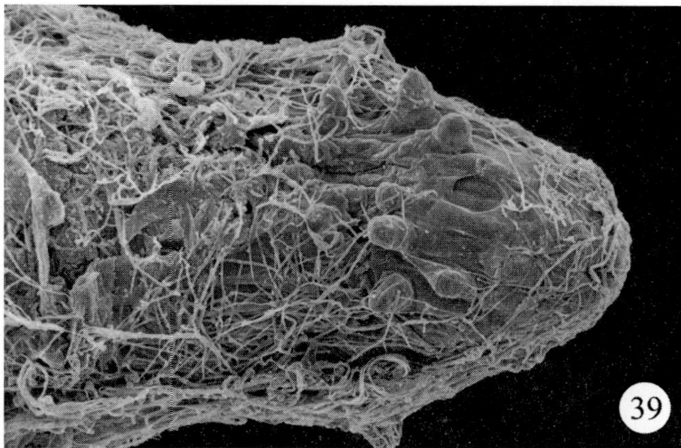
39

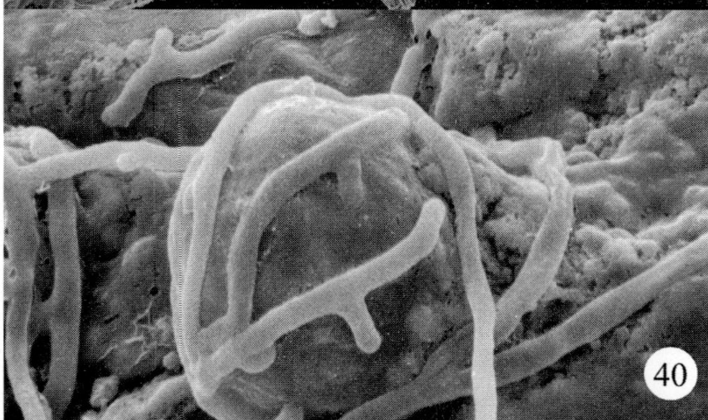
40

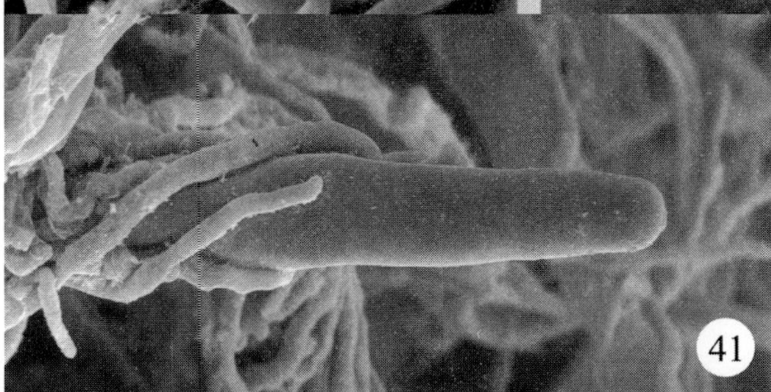
41

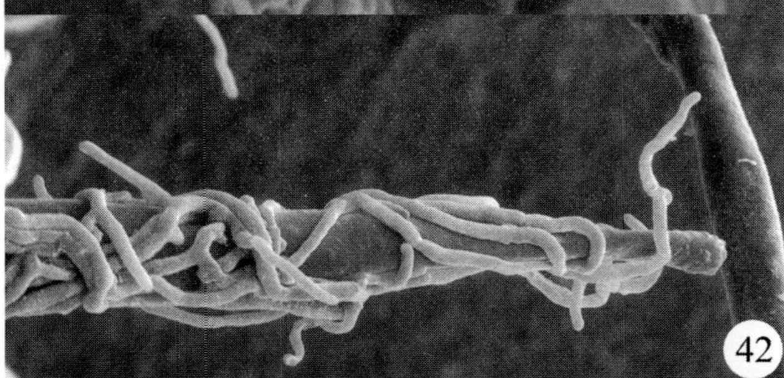
42

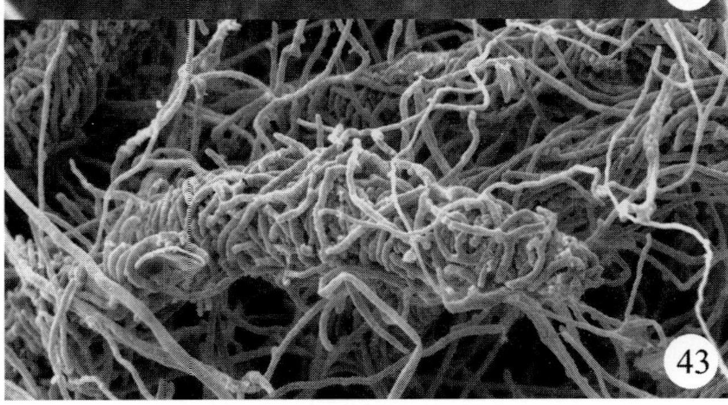
43

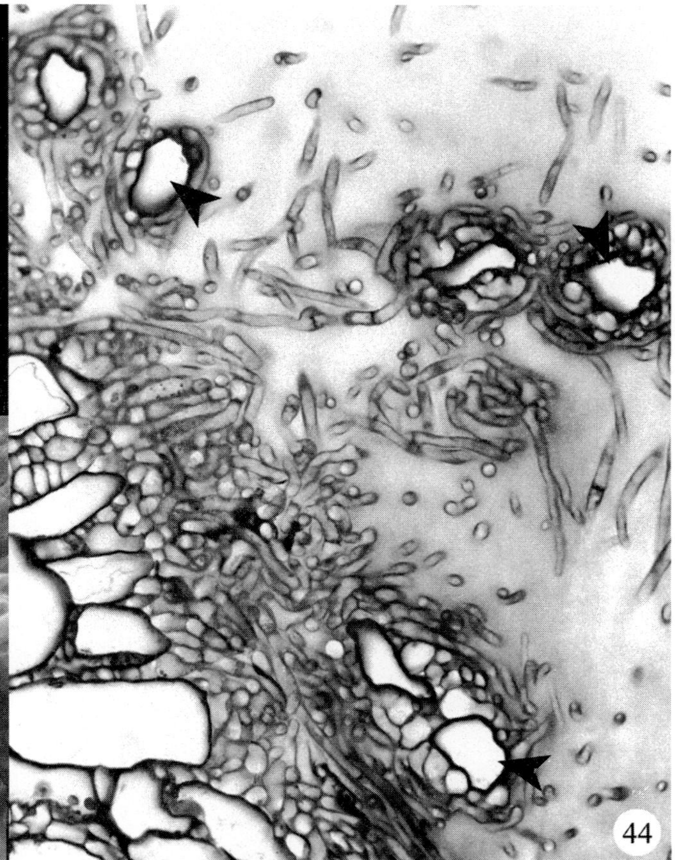
44

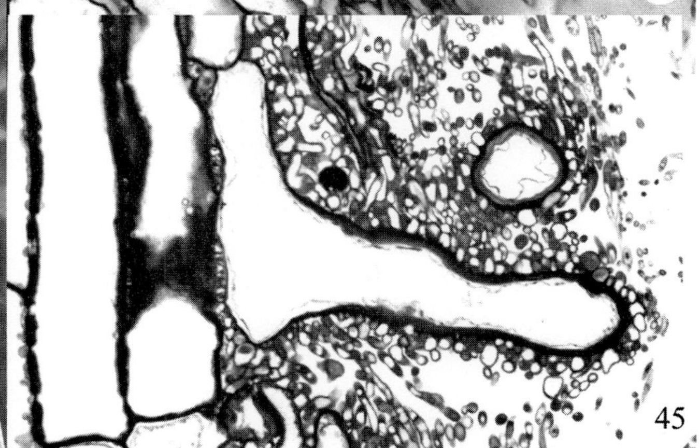
45

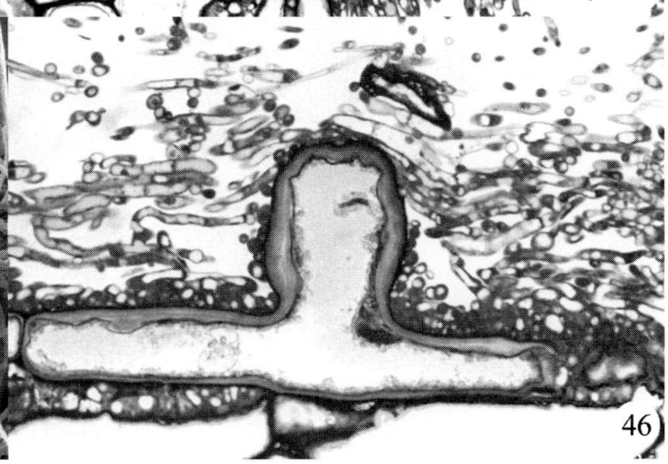
46

Figures 47–49. Mantle formation on short roots of *Pinus resinosa* colonized by *Cenococcum geophilum*.

Figure 50. Dichotomizing root tip of *Pinus resinosa* colonized by *Pisolithus tinctorius*. Several rhizomorphs (arrowheads) are evident.

Figure 51. Loose mantle of *Eucalyptus pilularis-Pisolithus tinctorius* ectomycorrhiza. Clamp connections (arrowheads) are visible.

Figures 52–55. Mantle of *Pinus ponderosa-Rhizopogon subcaerulescens* ectomycorrhiza viewed at increasing magnification. Numerous peg-like cystidia (arrowheads) occur in the outer mantle. From Massicotte et al. New Phytol. **142:** 355–370 (1999).

All figures are scanning electron micrographs.

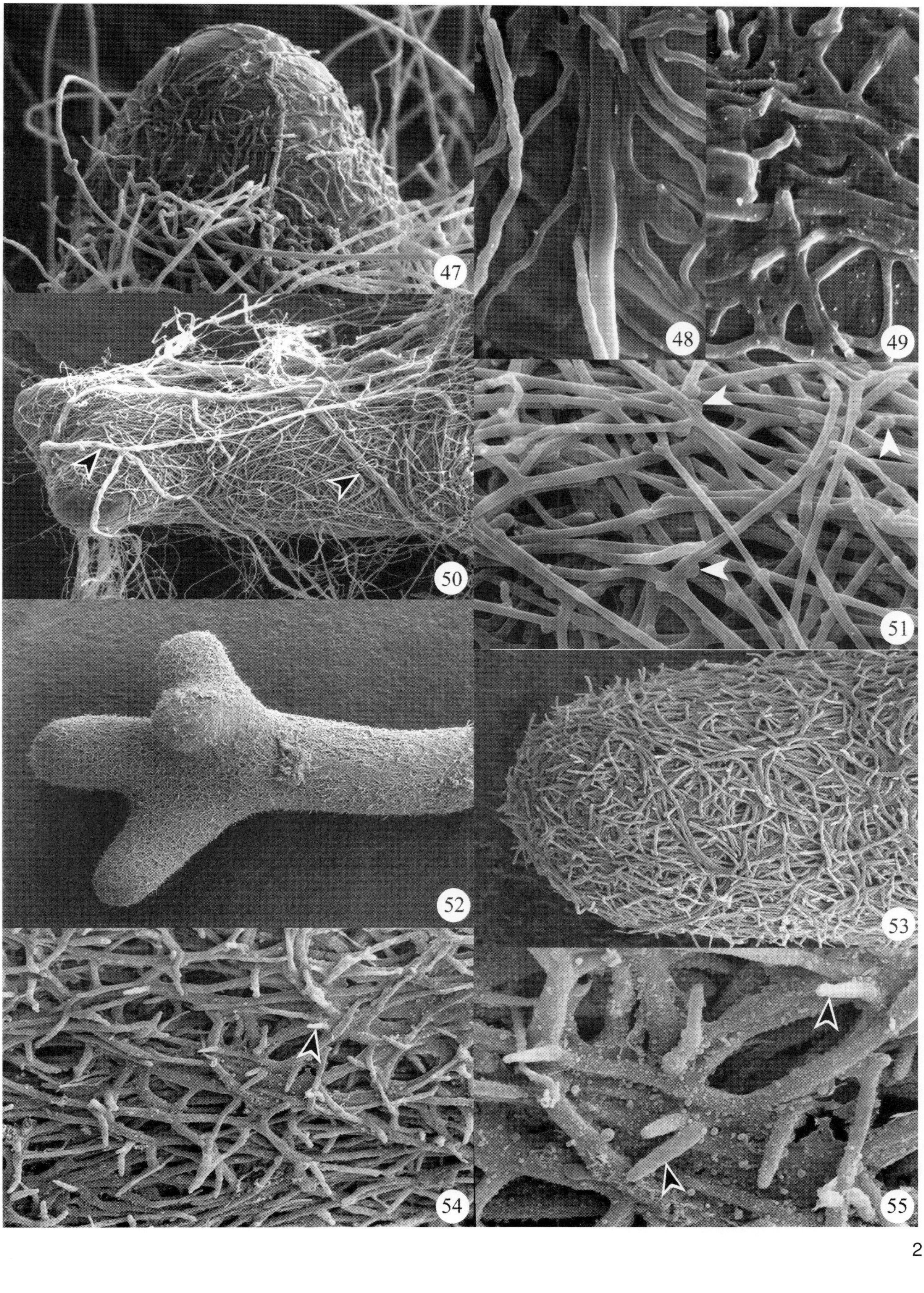
47
48
49
50
51
52
53
54
55

3. Functions

Although the main nutrient exchange interface in most ectomycorrhizas is the Hartig net, the repeated branching of inner mantle hyphae suggests that these may also be involved in bi-directional movement of nutrients. The fungus is capable of absorbing glucose and /or fructose from root cells and converting these into the soluble carbohydrates, trehalose and mannitol, or into the insoluble carbohydrate, glycogen (Figures 60, 61). These compounds may either be stored temporarily or for a longer term in mantle hyphae. In addition, mantle hyphae may accumulate other compounds including lipids (Figure 62), proteins, phenolics and polyphosphates (**see Box 1** for a discussion of polyphosphates). Deleterious metals may be bound to polyphosphates and other vacuolar deposits in the mantle, thereby preventing their uptake into roots. This observation is of particular relevance when polluted sites are being planted with tree seedlings inoculated with ectomycorrhizal fungi.

The compact nature of mantles of some ectomycorrhizas may contribute to protection of roots from water loss as soils dry and from ingress of pathogenic organisms. Loosely organized mantles probably have little impact on these processes. Since the mantle interfaces with the soil, it potentially affects the transport of water and nutrient ions into the root (**see Box 2**).

The presence of bacteria on the surface of and within mantles has been reported frequently. Bacteria may also be located within hyphae and root cells of some ectomycorrhizas (Nurmiaho-Lassila et al. 1997) but the function of these has not been determined. Some bacterial species located on the surface of mantle hyphae ("mycorrhiza helper bacteria") are known to enhance mycorrhiza formation and consequently plant performance (Garbaye 1994); others are detrimental to mycorrhiza development. Some bacteria associated with the mantle are able to fix atmospheric nitrogen, the fixation products of which can be utilized by the plant.

D. Hartig net

1. Development

The Hartig net, comprised of hyphae originating from the inner mantle, develops between root cells and forms a complex nutrient exchange interface over the surface of these cells. In most angiosperm species, the Hartig net develops only around epidermal cells; further development is blocked by the suberized walls of underlying exodermal cells (Figures 63, 64). One notable exception is the genus *Dryas,* where Hartig net hyphae develop around cells in much of the cortex; here Phi thickenings (lignified wall thickenings shaped like the Greek letter (Φ)) appear to block the ingress of the fungus (Melville et al. 1987). The Hartig net is shown in detail in longitudinal section (Figure 65) and in paradermal view (Figure 66). In the tree species *Pisonia grandis,* and perhaps a few other species, a Hartig net fails to develop. Here the inner mantle becomes the interface between the fungus and root epidermal cells. In *P. grandis,* epidermal cells and the few cortical cells that are in contact with fungal hyphae develop wall ingrowths surrounded by host plasma membrane, typical of transfer cells (Ashford and Allaway 1982). The development of ingrowths along the cell wall adjacent to fungal hyphae increases the surface area for nutrient exchange thereby replacing the Hartig net in this function (Allaway et al. 1985). Measurements of the surface area of cells with wall ingrowths compared to cells without indicate their potential as increased sites of nutrient exchange (**see Box 3**).

In conifers, the Hartig net develops around both epidermal and cortical cells and frequently occupies most of the volume of the cortex (Figures 67–70). During the early stage of Hartig net formation, the intrusion of hyphae by mechanical means into the middle lamella between epidermal and cortical cells is evident by the tapered nature of hyphal tips (Figures 71, 72). The process may be enhanced by the production of hydrolytic enzymes that soften the middle lamella and adjacent cell walls. With continued growth of the Hartig net hyphae, most of the intercellular connections (plasmodesmata) between root cells may be lost; however, in some ectomycorrhizas, groups of plasmodesmata remain (Figure 72).

The most pronounced feature of the Hartig net is its complex hyphal branching, often referred to as labyrinthine in nature, where tubular fungal hyphae are replaced by a multi-digitate mode of growth (Figure 73). This increases the surface area for exchange of nutrients. The cytoplasm of these highly branched hyphae is often enriched with organelles, including mitochondria, evidence for enhanced metabolic activity (Figure 73). Dolipore septa, typical of basidiomycetes, are frequent in Hartig net hyphae (Figure 74).

2. Functions

The Hartig net is involved in nutrient exchange in that it is through these hyphae that most of the sugars are absorbed by the fungus and mineral nutrients and water are passed to the root cells. Autoradiographic analysis has shown the pathway of sugars from root cells into Hartig net hyphae and then into the mantle (Duddridge et al. 1988; Bücking and Heyser 2001) and phosphate from the mantle into the Hartig net and eventually into the root cells (Bücking and Heyser 2001). In conifers, where the Hartig net can occupy much of the cortex, it is not clear whether cells in each zone of the cortex contribute equally to the transfer of nutrients. In most angiosperms, where the Hartig net is confined to the epidermis, the nutrient exchange interface is limited to this layer of cells. In both conifers and angiosperms, there is a gradient along the length of each mycorrhiza with the most active hyphae being located close to the root tip and senescence of hyphae occurring further back. Hartig net hyphae may also function as a depository for soluble and insoluble carbohydrates, lipids, phenolic compounds and polyphosphates.

E. Extraradical mycelium

1. Development

Hyphae that develop from the outer mantle into the surrounding soil are referred to as extraradical (= extramatrical) mycelium. Although this feature is difficult to study *in situ,* considerable information has been obtained observing the roots of plants grown in plexiglass chambers containing either soil from field sites or other substrates. It is evident that the extraradical mycelium can be an extensive network consisting of mycelial fans that permeate the soil and that may interconnect roots of the same plant and (or) other adjacent plants (Figure 75). The details of the development of these hyphae from the surface of the mantle into the soil are poorly documented. It is known, with respect to ectomycorrhizas formed with the fungal genus *Cenococcum,* that a certain number of hyphal layers are necessary in the mantle before the extraradical mycelium is initiated. The mycelium network often consists of hyphae and rhizomorphs (Figure 76), and in *Cenococcum* hyphae are thick-walled and pigmented (Figure 77). As extraradical hyphae develop they may adhere to and surround soil particles, or soil particles may adhere to the length of individual hyphae or groups of hyphae. Extraradical hyphae of some fungi grow around and into colonies of bacteria, excrement of earthworms, pollen, and organic debris in the soil; some are able to penetrate rock (**see Box 4**). Some fungi, particularly in the genera *Hysterangium* and *Gautieria,* form 'mats' of mycelium that bind soil and fine roots. This mycelium that may be encrusted with calcium oxalate crystals comprises a significant portion of the biomass in forest soils.

2. Rhizomorphs

Many ectomycorrhizal fungal species form complexes of extraradical hyphae (rhizomorphs or hyphal strands) that can vary considerably in their morphology, colour, and internal structure (Agerer 1992) (Figures 76, 78, 79). Each rhizomorph (strand) consists of a variable number of individual hyphae that interconnect with each other by different mechanisms. Although a number of terms are used to describe the variation in structure (Cairney et al. 1991), we will simply refer to these as rhizomorphs. Some rhizomorphs have calcium oxalate crystal ornamentations of specific size and shape on the surface hyphae and these, along with other features, are often used in species identification (Agerer 1987–2002). Sections through rhizomorphs reveal differences in anatomy. In the more complex rhizomorphs, one or more central hyphae (vessel hyphae) are enlarged (Figure 79) and these may have modified septa that allow for rapid movement of water and nutrient minerals (Cairney 1992). The peripheral hyphae of rhizomorphs may have thickened, pigmented walls that may be involved either in structural support or in providing resistance to the loss of water laterally.

Figure 56. Bacteria (arrowheads) present among mantle hyphae of a *Pinus strobus-Tuber* sp. ectomycorrhiza. Bacteria were stained with fluorescein isothiocyanate (FITC) and viewed with confocal laser scanning microscopy.

Figure 57. Bacteria (arrowhead) on mantle surface of a *Pinus resinosa-Laccaria bicolor* mycorrhiza viewed by scanning electron microscopy.

Figures 58–59. Longitudinal sections of *Betula alleghaniensis-Laccaria bicolor* ectomycorrhizas showing various degrees of branching (arrowheads) of Hartig net hyphae.

Figures 60–61. Ectomycorrhizas of *Eucalyptus pilularis-Hydnangium carneum* stained with acriflavine hydrochloride, a fluorescent probe for polysaccharides. Glycogen granules (arrowheads) are stored in mantle hyphae.

Figure 62. Transverse section of a *Fagus grandifolia* ectomycorrhiza (unknown fungal species) stained with fluorol yellow and examined by epifluorescence microscopy. Lipids (arrowheads) are present in mantle hyphae. From Herr and Peterson. Botanica Acta, **109**: 64-71. (1996)

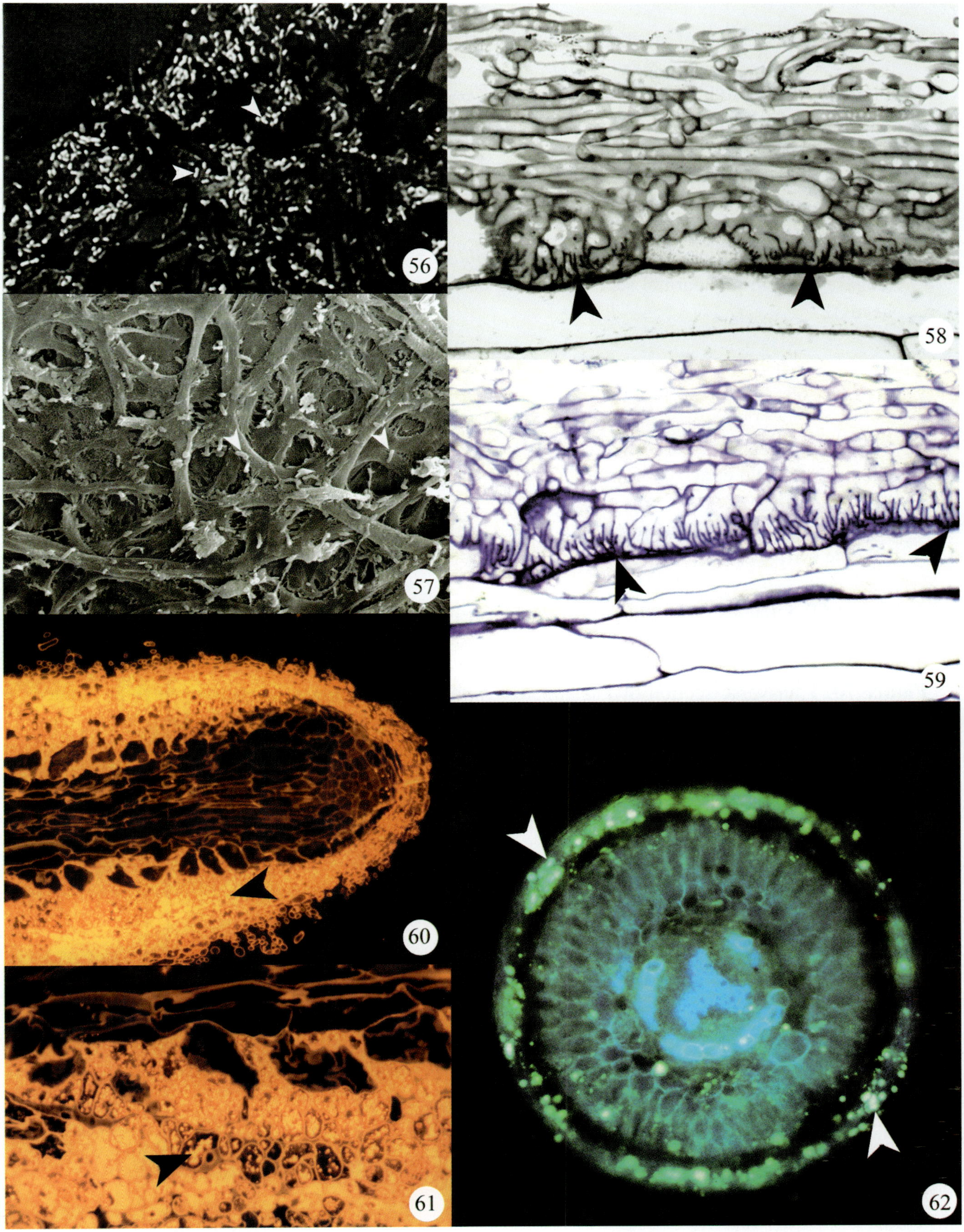
56
57
58
59
60
61
62

Figures 63–66. Sections of angiosperm ectomycorrhizas.

Figure 63. Longitudinal section of *Alnus crispa-Alpova diplophloeus* ectomycorrhiza showing the root meristem (*), the mantle (m), and paraepidermal Hartig net (arrowheads).

Figure 64. Transverse section of *Eucalyptus pilularis-Pisolithus tinctorius* ectomycorrhiza with a mantle (m) and paraepidermal Hartig net (arrowheads).

Figures 65, 66. *Alnus crispa-Alpova diplophloeus* ectomycorrhizas in longitudinal section (Figure 65) and paradermal section (Figure 66) showing the paraepidermal Hartig net (arrowheads).

Figures. 67–69. *Pinus resinosa-Piloderma bicolor* ectomycorrhizas.

Figure 67. Dichotomy of root apex. The mantle covers the entire dichotomous root.
From Massicotte et al. *Am. J. Bot.* **76**: 1654-1667. (1989)

Figure 68. Loose outer mantle (*), compact inner mantle (arrowheads) and Hartig net hyphae (double arrowheads).

Figure 69. Glancing section of Hartig net showing labyrinthine branching (arrowheads).

Figure 70. Transverse section of *Pseudotsuga menziesii-Rhizopogon* sp. ectomycorrhiza stained with sulforhodamine G and viewed with confocal laser scanning microscopy. Hartig net hyphae (arrowheads) are present around epidermal (e) and cortical (c) cells.

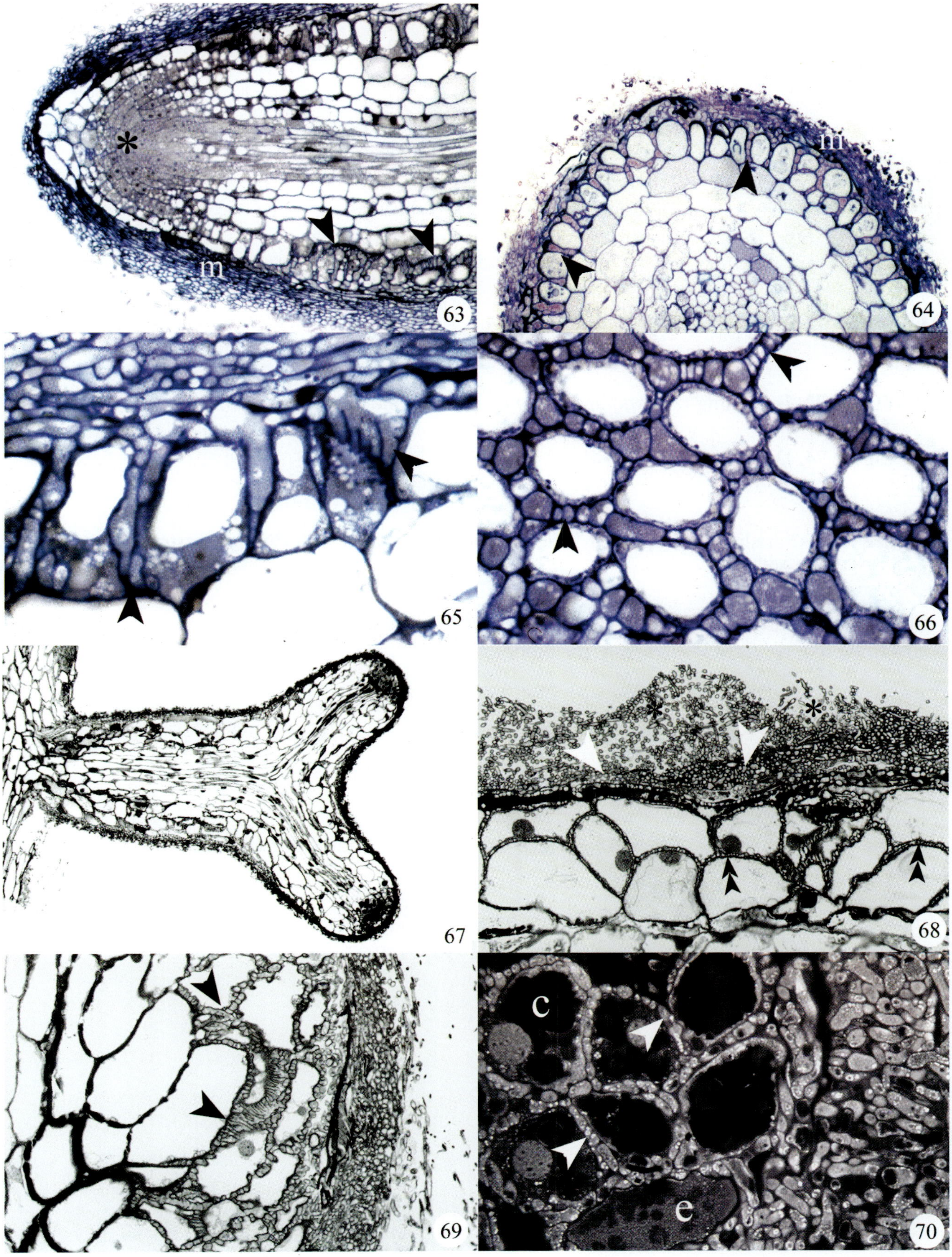
m
63
m
64
65
66
67
68
c
69
e
70

Figures 71, 72. Early stages in Hartig net formation in a *Dryas integrifolia-Hebeloma cylindrosporum* ectomycorrhiza.

Figure 71. A hyphal tip has become tapered (arrowhead) as it penetrates between epidermal cells (e).

Figure 72. The region between cortical cells (c) that has not been separated by Hartig net hyphae (box) has plasmodesmata (arrowheads in insert) present.

Figures 73, 74. Portions of Hartig net of a *Betula alleghaniensis-Pisolithus tinctorius* ectomycorrhiza.

Figure 73. Branching (arrows) and elongated mitochondria (arrowheads).

Figure 74. Dolipore septa (arrowheads).

Figure 75. Extraradical mycelium of *Suillus bovinus* associated with *Pinus sylvestris*. Photo courtesy of D. J. Read.

Figure 76. Extraradical hyphae of *Pisolithus tinctorius* associated with roots of *Pinus strobus*. Fine absorbing hyphae (arrowheads) and rhizomorphs (arrows) are present.

Figure 77. *Cenococcum* morphotype on *Abies lasiocarpa* (subalpine fir) exhibiting extensive emanating hyphae. Photo courtesy of L. Tackaberry.

Figure 78. Rhizomorphs of *Pisolithus tinctorius* associated with *Picea mariana*. (SEM).

Figure 79. Longitudinal section of a *Suillus pictus* rhizomorph showing enlarged 'vessel' hyphae. From: Randall and Grand. Can. J. Bot. **64:** 2182–2191 (1986).

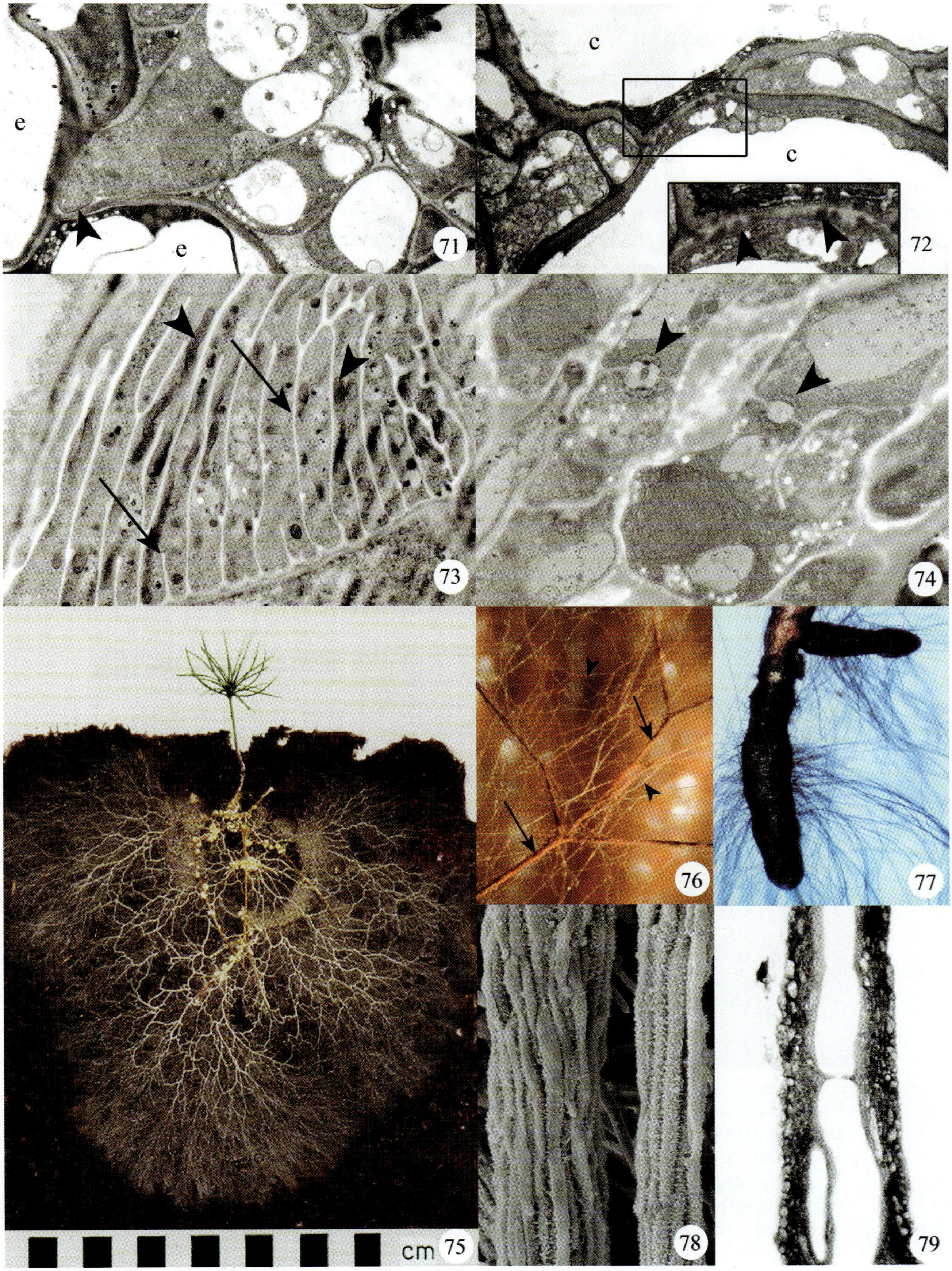
e
e
71
c
c
72
73
74
cm
75
76
77
78
79

Box 1: The polyphosphate controversy

The movement of phosphate from the soil solution through fungal hyphae to the plant root is one of the most important functions of ectomycorrhizal associations. The vacuolar polyphosphate granules frequently observed in mantle hyphae of ectomycorrhizas processed using conventional fixation and dehydration methods for light microscopy (Ashford et al. 1975), and transmission electron microscopy (Ashford et al. 1986; Strullu et al. 1982), have been interpreted as being artefactual since observations of living hyphae of *Pisolithus tinctorius* and hyphae processed for transmission electron microscopy by rapid-freezing followed by freeze substitution lack these granules (Orlovich and Ashford 1993). In *Eucalyptus pilularis-P. tinctorius* mycorrhizas, polyphosphate appears to be present in hyphal vacuoles in a dispersed form (Ashford et al. 1999). However, a recent study of *P. tinctorius* and three other ectomycorrhizal fungal species associated with *Pinus sylvestris* (Bücking and Heyser 1999) claims that polyphosphate granules are present in living hyphae and that these are retained during processing for microscopy. There is obviously scope for further research in this area. Hyphae of *P. tinctorius* possess a motile vacuole system (Shepherd et al. 1993*a,b* ; Ashford et al. 2001) that apparently can sequester and move phosphorus complexed with potassium (Cole et al. 1998). Extraradical and surface hyphae of fungal mantles of *E. pilularis-P. tinctorius* ectomycorrhizas also have a pleiomorphic vacuole system, a necessary requirement for a model to explain movement of phosphorus from the tips of absorbing extramatrical hyphae to the interface between the mantle and the root (Allaway and Ashford 2001). Further studies will undoubtedly resolve the important issue of phosphorus movement in mycorrhizal systems.

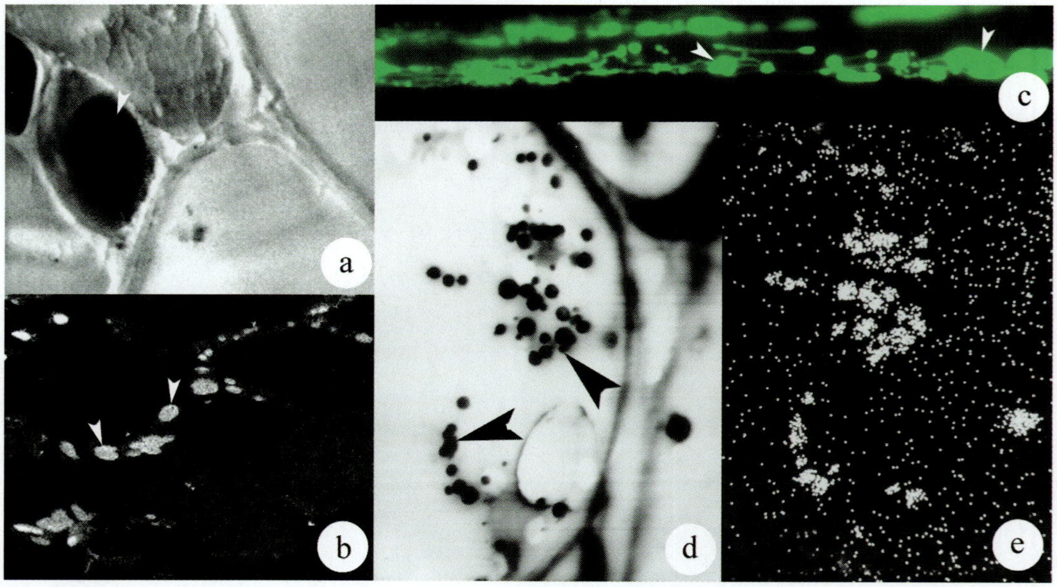

Figures a and b. Vacuoles (arrowheads) in Hartig net hyphae of *Eucalyptus pilularis-Pisolithus tinctorius* ectomycorrhiza. Tissue rapidly frozen and freeze substituted. In (b), X-ray map for phosphorus. From Ashford et al. Fungal Genetics and Biology **28:** 21–33 (1999). **Figure c**. Motile vacuole system in *Pisolithus tinctorius* hyphae labelled with carboxy-DFFDA. Photo courtesy of Dr. Anne Ashford. **Figures d and e**. Polyphosphate granules in mantle hyphae of an *Eucalyptus pilularis-Hydnangium carneum* ectomycorrhiza fixed by conventional methods. In (d) granules (arrowheads) as viewed with TEM. In (e) phosphorus dot map of the same granules in Fig (d) revealed by X-ray microanalysis. From Moore et al. New Phytol. **112:** 193–204 (1989).

Box 2: Are water and nutrients able to pass through the mantle?

The fungal mantle forms the interface between the root and the soil and could potentially play a role in absorption of water and nutrients. However, there is some debate as to whether water and mineral nutrients can pass through the mantle apoplastically (i.e., through the interhyphal spaces and hyphal cell walls without crossing the fungal plasma membranes). Some mantles appear to be impermeable to fluorescent probes (Vesk et al. 2000) while others are not (Behrmann and Heyser 1992); this is perhaps not surprising considering the diversity of mantle types that comprise the numerous plant/fungus combinations. Lanthanum, a tracer that is transported apoplastically and that may be a good indicator of ion movement, does penetrate the mantle and is only blocked by Casparian bands or suberin lamellae of exodermal and endodermal cells (Vesk et al. 2000; Behrmann and Heyser 1992). An understanding of the permeability of the mantle is important when considering the function of this component of an ectomycorrhiza. One must also remember that most of the ions being transported to the root by ectomycorrhizal fungi are taken up by the extraradical mycelium at some distance from the mantle surface and must be transported to the mantle and subsequently to the Hartig net hyphae before they can be taken up by the root cells.

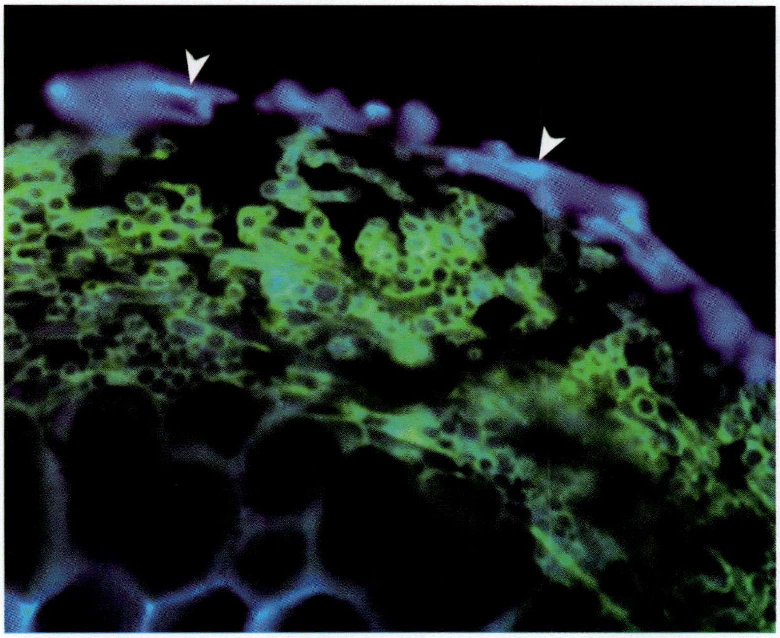

Eucalyptus pilularis-Pisolithus tinctorius ectomycorrhiza showing restriction of the fluorescent probe, PTS, to the outer mantle (arrowheads). From Vesk et al. New Phytol. **145**: 333–346 (2000).

Box 3: Root cells may respond to mycorrhizal fungi by developing as transfer cells

The exchange of nutrients between plant and fungus in ectomycorrhizas is complex. Signalling between the symbionts results in structural changes in both. One manifestation of this is the response of host cells in some mycorrhizas to develop transfer cells. Transfer cells are parenchyma cells that become modified by the elaboration of wall ingrowths and associated plasma membrane providing an increased surface area for short distance transport of nutrients (Gunning and Pate 1974). The first report of transfer cells being induced by ectomycorrhizal fungi (Ashford and Allaway 1982) was described for ectomycorrhizas of the tree species, *Pisonia grandis,* occurring on two islands in the Australian Great Barrier Reef. In these ectomycorrhizas, a Hartig net fails to develop and ingrowths form along the epidermal cell wall that is in contact with the inner mantle. Later collections of *P. grandis* from the Seychelle Islands showed similar development, however, a Hartig net did form and wall ingrowths occurred where epidermal and cortical cells interfaced with the Hartig net (Ashford and Allaway 1985). A basidiomycete fungus isolated from roots of *P. grandis* was able to induce transfer cells in this tree species but not in the conifer, *Picea sitchensis* (Cairney et al. 1994), indicating that the genome of the plant species may have some control over transfer cell formation. Transfer cells also occur at the host–fungus interface in *Pinus sylvestris–Suillus bovinus* ectomycorrhizas (Duddridge and Read 1984), in tuberculate mycorrhizas of *Castanopsis borneensis* and *Engelhardtia roxburghiana* (Haug et al. 1991) and in *Alnus crispa–Alpova diplophloeus* ectomycorrhizas (Massicotte et al. 1986). In the latter, the composition of the wall ingrowths is different from the adjacent cell wall (Massicotte et al. 1987) and this may reflect an increase in permeability of wall ingrowths.

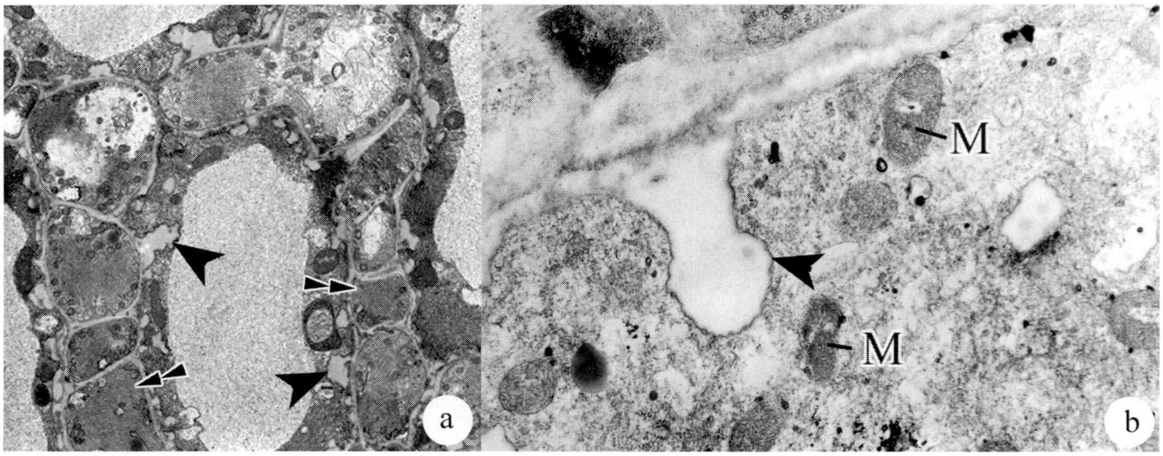

Figures a and b. *Alnus crispa–Alpova diplophloeus* ectomycorrhiza showing wall ingrowths (arrowheads) in epidermal cells adjacent to Hartig net hyphae (double arrowheads). Mitochondria (M) occur adjacent to the ingrowths.

3. Sclerotia

Relatively few ectomycorrhizal fungal species form sclerotia in the extraradical mycelium.The development and structure of sclerotia have been studied in detail for a few species: *Pisolithus tinctorius, Cenococcum geophilum, Hebeloma sacchariolens,* and *Paxillus involutus*. Other mycorrhizal fungi including species of *Gyrodon, Boletus, Austropaxillus, Cortinarius*, and perhaps *Morchella,* are also known to form these structures.

Sclerotia are initiated in the extraradical mycelium, and are sometimes associated with rhizomorphs (Figures 80, 81). Hyphae branch, come into contact with each other, and form discrete structures. At maturity, each sclerotium usually develops a melanized outer covering, the rind (Figure 82) that surrounds a cortex (central area) of compact hyphae, and a medulla of loosely organized hyphae. The cortical region stores proteins (Figure 83), lipids, polysaccharides (Figure 84), and polyphosphates, making these structures ideal propagules. Large numbers of *C. geophilum* sclerotia have been isolated from soil and from run-off water from forest soils. *C. geophilum* sclerotia produced under laboratory conditions develop thick-walled hyphae that emanate from the surface (Figures 85–87). Many sclerotia, including those formed by *Pisolithus,* can be stored under laboratory conditions for prolonged periods of time and then germinated under suitable conditions (Figure 88). Hyphae developing from germinated sclerotia are effective in forming typical ectomycorrhizas. Sclerotia as well as cultured mycelium can be used for long-term maintenance of fungal genotypes **(see Box 5)**.

4. Sexual reproductive structures

The formation of reproductive bodies, both basidiocarps and ascocarps, involves the localized branching of extraradical hyphae (Figure 89), the organization of these hyphae into discrete structures (Figures 90–92), and the differentiation of the various regions of the basidiocarp (Figure 93). At maturity, a reproductive structure typical of the fungal species is formed (Figure 94). Basidiospores in the basidiomycetes, and ascospores in the ascomycetes formed in the reproductive structures, are dispersed by various mechanisms as propagules for the fungus.

5. Functions

The extraradical mycelium, highly variable in structure, has a significant role in ectomycorrhiza functioning. The most obvious function of the fine hyphae that comprise much of the extraradical mycelium is the mobilization, absorption and translocation of mineral nutrients and water from the soil substrate to plant roots. In those species with rhizomorphs, connecting fine hyphae may pass water and dissolved nutrients to these structures for more rapid translocation through the wide diameter central hyphae (vessel hyphae) to the root. At some point, the extraradical hyphae and rhizomorphs interface with the mantle where transfer must occur to outer mantle hyphae. Experiments with radioactive isotopes of phosphorus (^{13}P-labelled orthophosphate) have shown that phosphorus can be translocated over distances of more than 40 cm through rhizomorphs to roots of colonized plants and subsequently to the shoot system (Finlay and Read 1986*b*). These same authors provide further evidence of the importance of these hyphae in the uptake of phosphorus by showing that the fine hyphae of mycelial fans become heavily labelled when ^{32}P-labelled orthophosphate is added to the soil. The demonstration that ectomycorrhizal fungi can obtain phosphorus from the mycelium network of a saprotrophic fungus and pass phosphorus to the host plant (Lindahl et al. 1999) illustrates the dynamics occurring in the rhizosphere of ectomycorrhizas and the complexity of below-ground processes.

Carbon compounds are translocated in the reverse direction from the host root to the extraradical mycelium for metabolic and growth processes, to developing sclerotia and their storage reserves, and to sexual reproductive bodies (e.g., basidiocarps and ascocarps) that act as major sinks for host-derived carbon. In years in which numerous basidiocarps form, this can be a carbon drain on the host tree. Limited experimental evidence shows that the rate of production and final biomass of *Laccaria bicolor* basidiocarps is correlated with the rate of photosynthesis of their host, *Pinus strobus* (Lamhamedi et al. 1994).

Figures 80–82. Developing sclerotia from rhizomorphs (arrowhead) of *Pisolithus tinctorius* associated with roots of *Pinus strobus*. The sclerotium in Figure 82 has a melanized rind.

Figures 83–84. Sections through sclerotia of *Paxillus involutus* stained for proteins (arrows) (Figure 83) and polysaccharides (yellow granules) (Figure 84).

Figures 85–87. Sclerotia of *Cenococcum geophilum* produced under laboratory conditions showing the characteristic coarse hyphae (arrowheads) emanating from the rind. Figures 86 and 87 are scanning electron micrographs.

Figure 88. A germinating sclerotium of *Pisolithus tinctorius* on the surface of a nutrient medium.

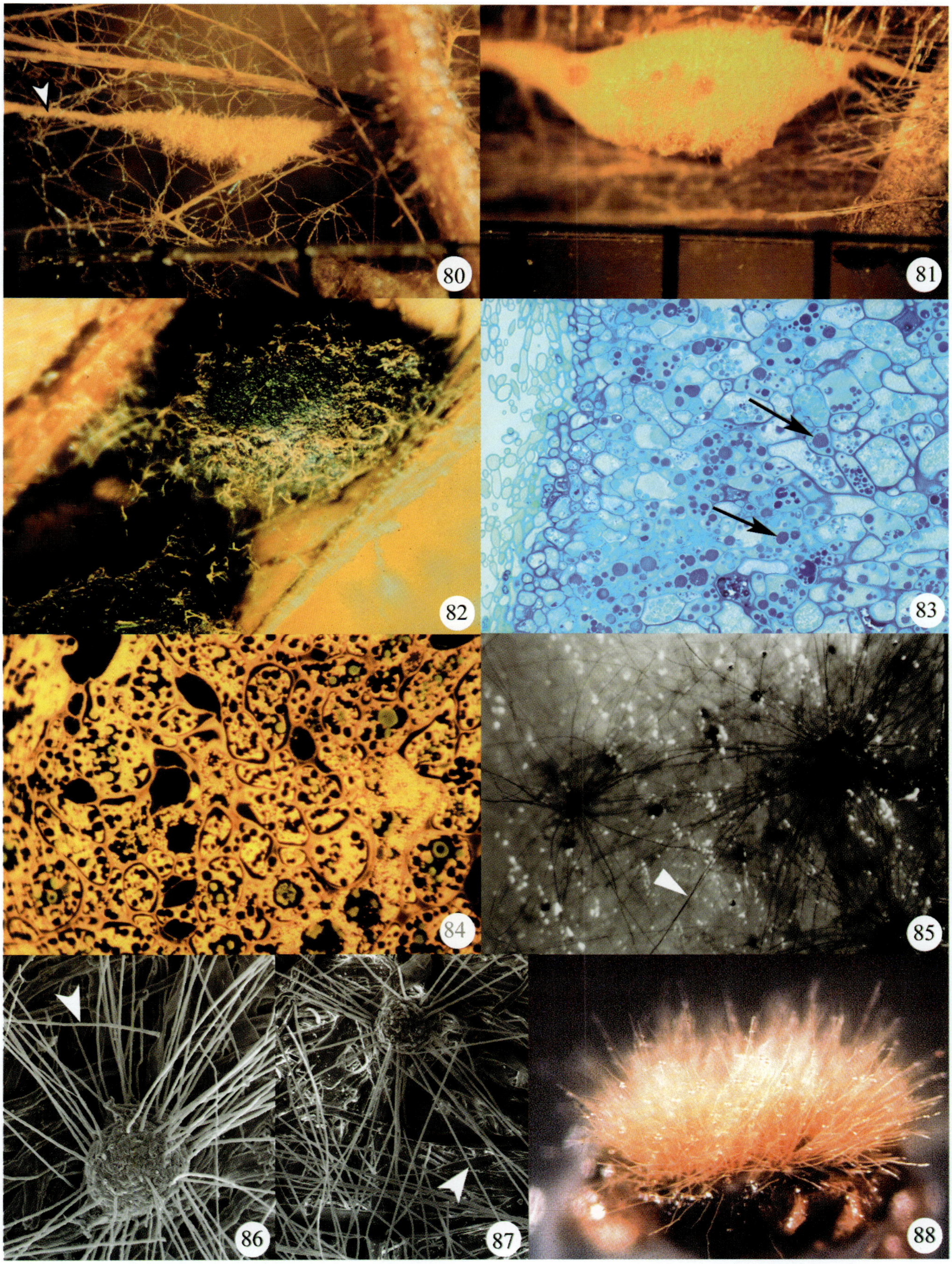
80
81
82
83
84
85
86
87
88

Figures 89–93. Developing basidiocarps of *Laccaria bicolor* in the extraradical mycelium associated with roots of *Pinus resinosa*.

Figure 89. Aggregated and branched hyphae initiating a basidiocarp primordium. (Scanning electron microscopy).

Figure 90. Young basidiocarp primordium. (Light microscopy).

Figures 91, 92. Basidiocarp primordia differentiated into a stipe and pileus region. Figure 91. (Light microscopy). Figure 92. (Scanning electron microscopy).

Figure 93. In sectional view, the hyphae of the pileus region assume a pseudoparenchymatous nature.

Figure 94. A basidiocarp of *Laccaria laccata* formed in a pot culture with *Quercus acutissima*.

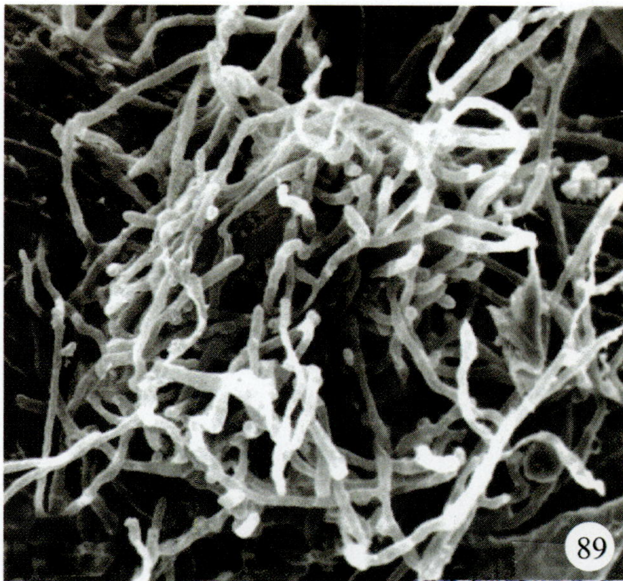
89

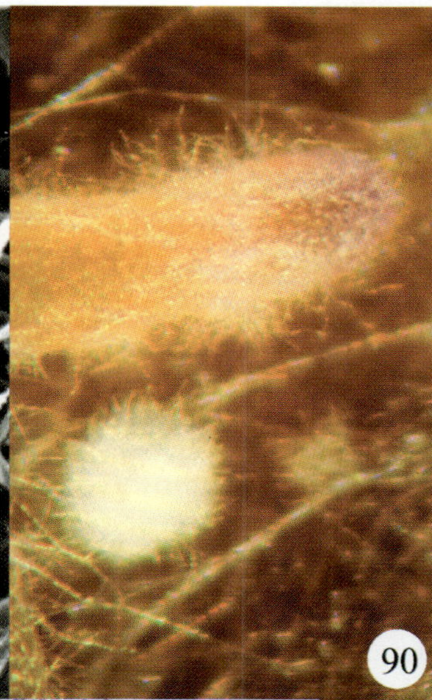
90

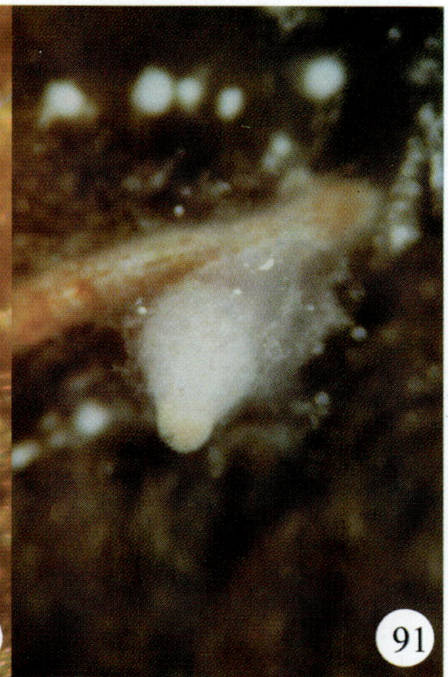
91

92

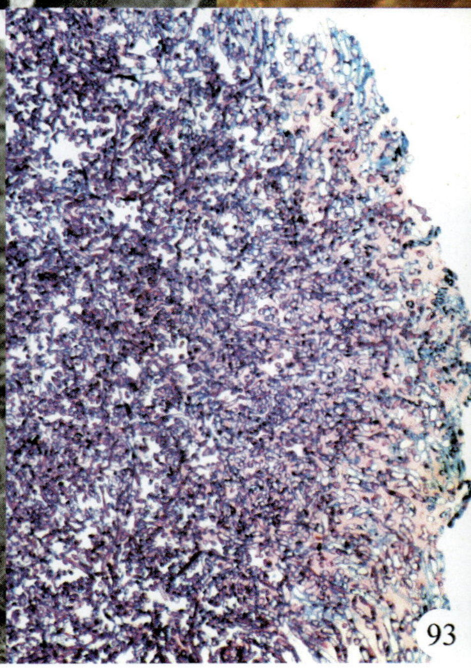
93

94

Experiments with radioisotopes of carbon in the laboratory have confirmed the movement of carbon from host to fungus as well as from one plant to another through the extraradical mycelium network (Finlay and Read 1986*a*). There is increasing evidence in the field that nutrients can be translocated from one plant to another through hyphal links and the ecological significance of this is being explored.

Hyphae in fungal 'mats' secrete various organic acids that appear to be involved in the weathering of mineral soils thereby releasing recalcitrant minerals that can be taken up by the mycelium and translocated to roots. This is an exciting current line of research (**see Box 4**).

Bacteria are known to associate with extraradical hyphae to form 'biofilms' (layers of bacteria embedded in secreted polysaccharides). Many of the bacteria isolated from these biofilms are able to break down petroleum hydrocarbons and other soil pollutants (Sarand et al. 1998). These authors showed that the mycelium of the ectomycorrhizal fungus *Suillus bovinus* and associated biofilms of bacteria that emanated from seedling roots of *Pinus sylvestris* proliferated in patches of soil contaminated with petroleum hydrocarbons (Figure 95). The planting of tree seedlings with associated mycorrhizal fungi and appropriate bacteria may become important in restoration projects aimed at reclaiming soils polluted with various industrial by-products, such as petroleum hydrocarbons and heavy metals.

Figure 95. *Pinus sylvestris* seedling colonized by the ectomycorrhizal fungus *Suillus bovinus*. A patch of extraradical mycelium (arrowheads) has proliferated over introduced petroleum hydrocarbon contaminated soil.

Photo from Sarand et al. FEMS Microbiol. Ecol. **27:** 115–126 (1998).

F. Specialized ectomycorrhizas – tuberculate mycorrhizas

When associated with certain fungal species, several conifers, including some *Pinus* species (Randall and Grand 1986), *Pseudotsuga* (Massicotte et al. 1992), and *Tsuga* as well as several angiosperms such as *Eucalyptus* (Dell et al. 1990), *Quercus, Castanopsis* and others, develop clusters of lateral roots known as tubercles (Figures 96–98). Each lateral root in the cluster develops a mantle (Figures 99–102) and Hartig net (Figures 103, 104), while the entire cluster of roots becomes covered, to a variable degree by compact layers of hyphae called the peridium or rind (Figures 99, 100). Although these structures are fairly common with some host species and in certain substrates, there is little information concerning their function. Hartig net hyphae and inner mantle hyphae may store polysaccharides and proteins. The suggestion has been made that the rind may protect the enclosed ectomycorrhizas from pathogens and insects. Bacteria that have been isolated from the surface and within the peridium of *Pseudotsuga* tubercles are able to fix atmospheric nitrogen; it is not known whether this nitrogen is utilized by the host.

Box 4: Can ectomycorrhizal fungi really eat rocks?

Introductory mycology courses teach us that fungi, in general, can be incredibly versatile when it comes to their diet, and it seems that ectomycorrhizal fungi are no exception: some apparently can eat rocks! Scientists from The Netherlands and Sweden documented (unearthed!) the fact that some ectomycorrhizal fungi are able to weather feldspars and hornblendes in Podzol horizons and granitic bedrock. Small pores are created by organic acids released from hyphae; this process results in the release of minerals that can be taken up by hyphae and translocated directly to host plants (Jongmans et al. 1997). These scientists estimated that, at any one time, millions of hyphal tips of *Piloderma croceum* and *Suillus granulatus* can eat their way through sand grains, thus participating in the process of nutrient acquisition. Other elemental "food sources" that ectomycorrhizal fungi are able to mobilize include phosphorus extracted from apatite (Wallander et al. 1997) as well as potassium extracted from biotite (Wallander and Wickman 1999; Glowa et al. 2003), microcline (Wallander and Wickman 1999) and phlogopite (Leyval et al. 1990; Paris et al. 1995). Arocena et al. (1999) have shown that mycorrhizal fungi can also extract magnesium from mica. Some researchers suggest that a synergy may exist between fungi and bacteria in weathering minerals (Leyval et al. 1990; Olsson and Wallander 1998). Several hypotheses have been proposed to explain how fungi can accomplish this "feast". The interaction between fungal symbionts and roots may contribute to reducing the pH, thereby increasing the solubility of minerals. In addition, it is known that both fungi and roots can produce organic acids (i.e., succinate, citrate, oxalate) which can dissolve minerals. Finally, a lowering of element concentration in the soil solution (thereby augmenting the concentration gradient) may increase the release of elements from mineral substrates.

Research continues in this area, however, it is clear that over time, not even rocks are spared from the patient work of fungi, thus contributing to tree nutrition and growth.

Figure a. A fungal 'mat' in the litter – root layer growing on rocks under a mature *Pinus sylvestris* forest near Uppsala, Sweden. **Figure b.** A network of mycelium and rhizomorphs (arrowheads) in a portion of the fungal 'mat' in Figure a.

Figure 96. Tuberculate cluster of an unknown fungal species from a soil core sample. Host is likely *Pinus strobus* (white pine).

Figure 97. Tuberculate cluster of an unknown fungal species on *Pinus contorta* (lodgepole pine).

Figures 98–104. Tuberculate ectomycorrhizas formed on root systems of *Pseudotsuga menziesii* collected from litter and rotting logs.

Figure 98. Several tuberculate clusters (*) along excavated roots.

Figure 99. Sectioned tuberculate cluster showing many lateral roots (arrows), fungal mycelium (arrowheads) and a sheath (peridium) (*).

Figure 100. Tuberculate cluster similar to that shown in Figure 99 viewed with scanning electron microscopy. Lateral roots (arrows) are sectioned transversely. The star indicates the sheath.

Figure 101. Tuberculate cluster with the sheath removed showing many lateral roots surrounded by a mantle (arrowheads). Scanning electron microscopy.

Figure 102. Enlargement of lateral roots, each with a mantle (arrowheads). Scanning electron microscopy.

Figure 103. Portion of one lateral root, embedded in LR White and sectioned, showing the mantle (arrow) and Hartig net hyphae (arrowheads) surrounding epidermal and cortical cells. A loosely organized mycelium (*) occurs next to the mantle.

Figure 104. Portions of two lateral roots from a tuberculate cluster, each with a Hartig net (arrowheads) and contiguous mantles (*).

Figures 100-102 from Massicotte et al. Trees, **6**: 137-146. (1992)
Material for Figures 103, 104 embedded in LR White and sectioned.

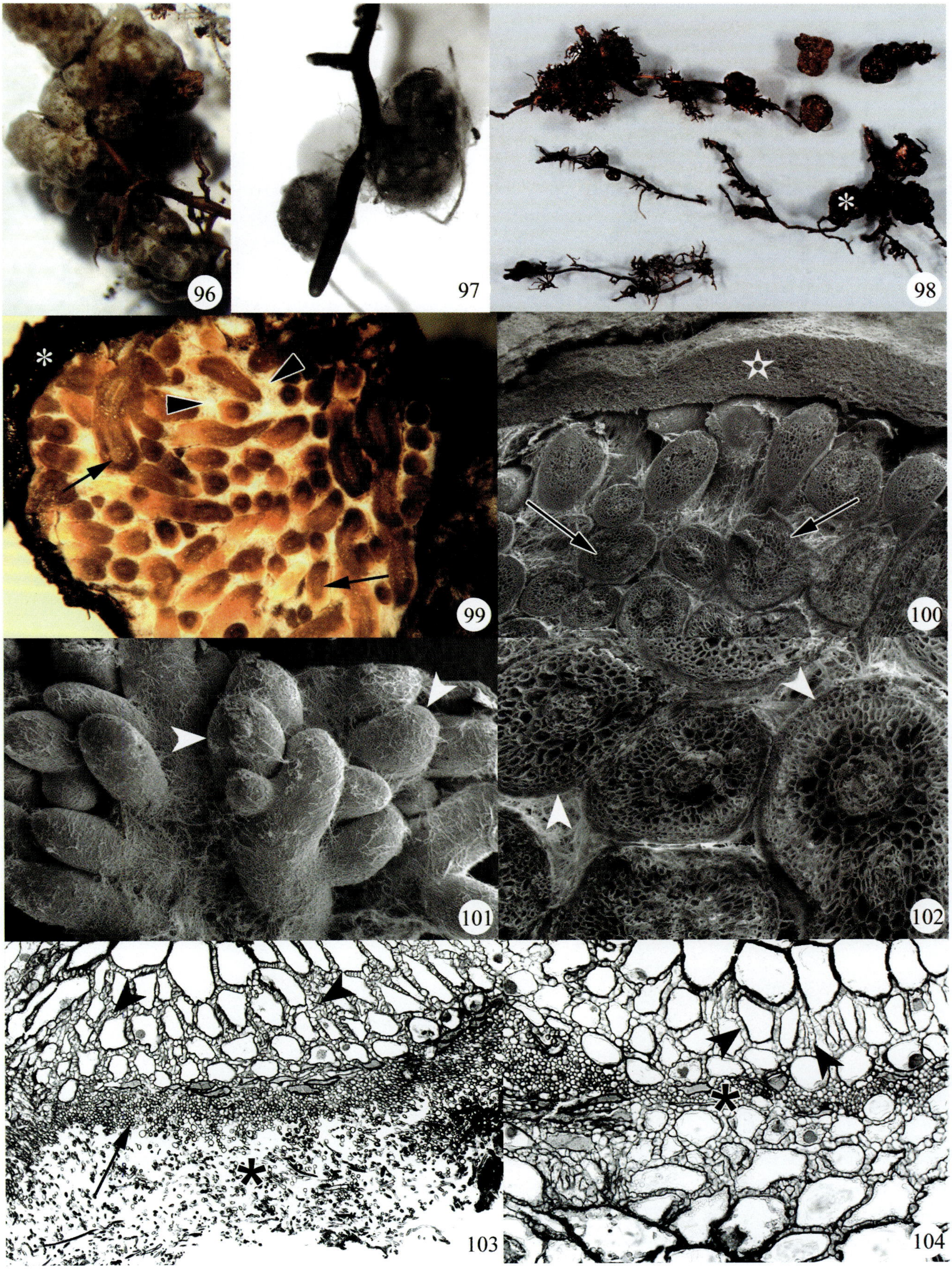
96
97
98
99
100
101
102
103
104

Box 5: Growing ectomycorrhizal fungi in the lab

Many fungal species that form ectomycorrhizas can be cultured on nutrient media under axenic conditions. Various sources of material can be used to establish cultures. Most frequently, pieces of axenic or sterile tissue from fungal fruitbodies are excised and placed on the surface of a nutrient medium, either in small vials or in Petri dishes. After hyphae grow from this tissue, subcultures are made to ensure that cultures are not contaminated by bacteria or other fungi. Cultures can also be started directly from ectomycorrhizal roots or from spores produced by fruitbodies. Cultures free of contaminants can be kept in long-term cold storage. Details of the methods used in obtaining sterile cultures of ectomycorrhizal fungi and maintaining these under long-term storage can be found in Brundrett et al. (1996).

Why establish sterile cultures of ectomycorrhizal fungi?

1. The mycelium produced can be used as a source of inoculum to test the ability of many fungal species to establish ectomycorrhizas with a variety of host plants.
2. Features of the mycelium can be used as an aid in the identification of fungal species, particularly if the cultures are initiated from ectomycorrhizal roots.
3. Experiments can be set up to determine the ability of fungal isolates to utilize various substrates, to determine the optimal temperature and pH conditions for growth, and to assess fungal performance when challenged with substances such as heavy metals or other environmental pollutants.
4. Experiments can be designed to determine the interaction between ectomycorrhizal fungi and other organisms, for example pathogenic fungi and bacteria.
5. Some fungal species can be grown in liquid culture and in fermentors to produce inoculum for field studies.
6. Axenic cultures provide material for long-term storage of fungal genotypes.

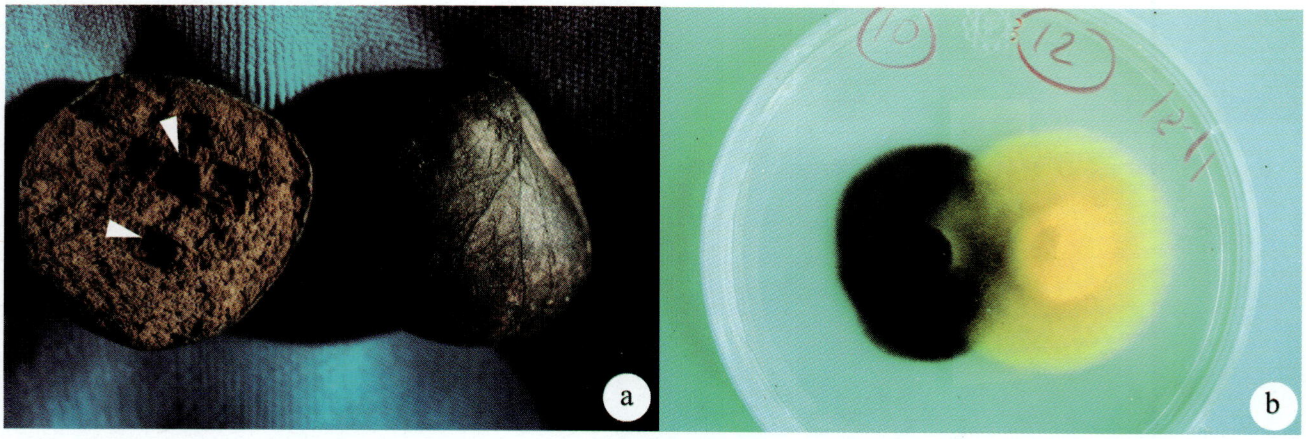

Figure a. Hypogeous fruit body of a *Rhizopogon* sp. showing where tissue has been removed for culture (arrowheads).
Figure b. Co-cultures of *Cenococcum geophilum* (black) and *Piloderma bicolor* (yellow).

Chapter 2. Ectendomycorrhizas

Pinus strobus forest
Northern Ontario
Canada
Exposed *Pinus ponderosa* roots
Southern British Columbia
Canada
Mixed conifer and hardwood
clearcut
S.W. Oregon
USA
Larix laricina
(tamarack)
Central British Columbia
Canada
St. Williams *Pinus strobus*
seedling nursery
Southern Ontario
Canada

Chapter 2. Ectendomycorrhizas

Introduction

1. Definition

Ectendomycorrhizas are associations formed between a limited number of ascomycete fungi and the conifer genera, *Pinus* and *Larix* (Yu et al. 2001*a*). Ectendomycorrhizas have been defined by others in various ways and sometimes have encompassed the arbutoid and monotropoid mycorrhizas. These two categories are described separately in this book. Structurally, ectendomycorrhizas resemble ectomycorrhizas; they have a mantle and Hartig net but differ in that after Hartig net formation, intracellular hyphae develop in epidermal and cortical cells (Figures 105–107).

2. Plant species involved

By our strict definition, only two conifer genera (*Pinus* and *Larix*), both in the Pinaceae, form true ectendomycorrhizas. Reports of other genera having this category of mycorrhiza are mainly based on field-collected samples and these may represent senescent ectomycorrhizas. Although the occurrence of ectendomycorrhizas is currently restricted to two genera of conifers, this represents a significant distribution in that there are close to 100 species of *Pinus* and 10 to12 species of *Larix*. However, few species of either genus have been investigated either in the field or under laboratory conditions for the occurrence of ectendomycorrhizas.

3. Fungal species involved

Initially the fungi forming ectendomycorrhizas were grouped as E-strain fungi mainly because sexual stages were not identified; only general morphological characteristics of hyphae and chlamydospores were used to characterize the isolates (Figures 108, 109). With the discovery of sexual stages and with the use of molecular methods, a limited number of ascomycetes have been identified as fungal partners in ectendomycorrhizas. Most of the isolates of ectendomycorrhizal fungi belong to two species of *Wilcoxina, W. mikolae* var. *mikolae* (Figure 110), *W. mikolae* var. *tetraspora* and *W. rehmii* in the ascomycete order, Pezizales (Egger 1996). Another member of the Pezizales (*Sphaerosporella brunnea)* forms ectendomycorrhizas with *Pinus contorta*. Two genera in the Leotiales (*Phialophora finlandia* and *Chloridium paucisporum)* form mycorrhizas resembling ectendomycorrhizas (Yu et al. 2001*a*). An interesting feature of several of these fungal species is that they are able to form typical ectomycorrhizas with a number of conifer and angiosperm species (Scales and Peterson 1991*a*). Therefore, although few fungal species are involved, many tree species could potentially form ectomycorrhizas with them (Yu et al. 2001*a*).

B. Structural characteristics

In *Pinus* species, ectendomycorrhizal fungi induce the typical dichotomy of short roots (Figure 111) that are also observed with ectomycorrhizal fungi; as well, clusters of dichotomous short roots often develop (Figure 112). Few detailed studies of the structure of ectendomycorrhizas have been published, the most thorough being that of Scales and Peterson (1991*b*) with two species of *Wilcoxina* in association with *Pinus banksiana*, grown under laboratory conditions. In this and other studies, the main features of ectendomycorrhizas are the colonization of short roots, the development of a thin mantle, a Hartig net, and intracellular hyphae (Figures 105–107). The thin mantle may not be apparent when roots are viewed at low magnification (Figure 113), but at higher magnification, mantle features, such as branching of hyphae can be seen (Figure 114). In early stages of mantle formation, hyphae may surround root hairs (Figures 115, 116) while others can be seen embedded in mucilage on the root surface (Figures 117, 118). Extraradical hyphae are not usually abundant. In sectional view, the mantle appears as a thin layer of hyphae partially embedded in mucilage (Figures 119, 120).

The Hartig net develops as a uniseriate layer between epidermal and cortical cells (Figures 119–121). Frequently, most of the cortex is occupied by Hartig net hyphae (Figures 121, 122).

Figures 105–107. Diagrammatic representations of ectendomycorrhizas.

Figure 105. Longitudinal section of root tip showing apical meristem (*), Hartig net (arrows), and intracellular hyphae (arrowheads).

Figure 106. Transverse section showing thin mantle (M), Hartig net hyphae (arrows), and intracellular hyphae (arrowheads). Cortical cell nuclei (N) occur surrounded by intracellular hyphae.

Figure 107. Diagram showing the relationship between Hartig net hyphae (arrows), and intracellular hyphae (arrowheads).

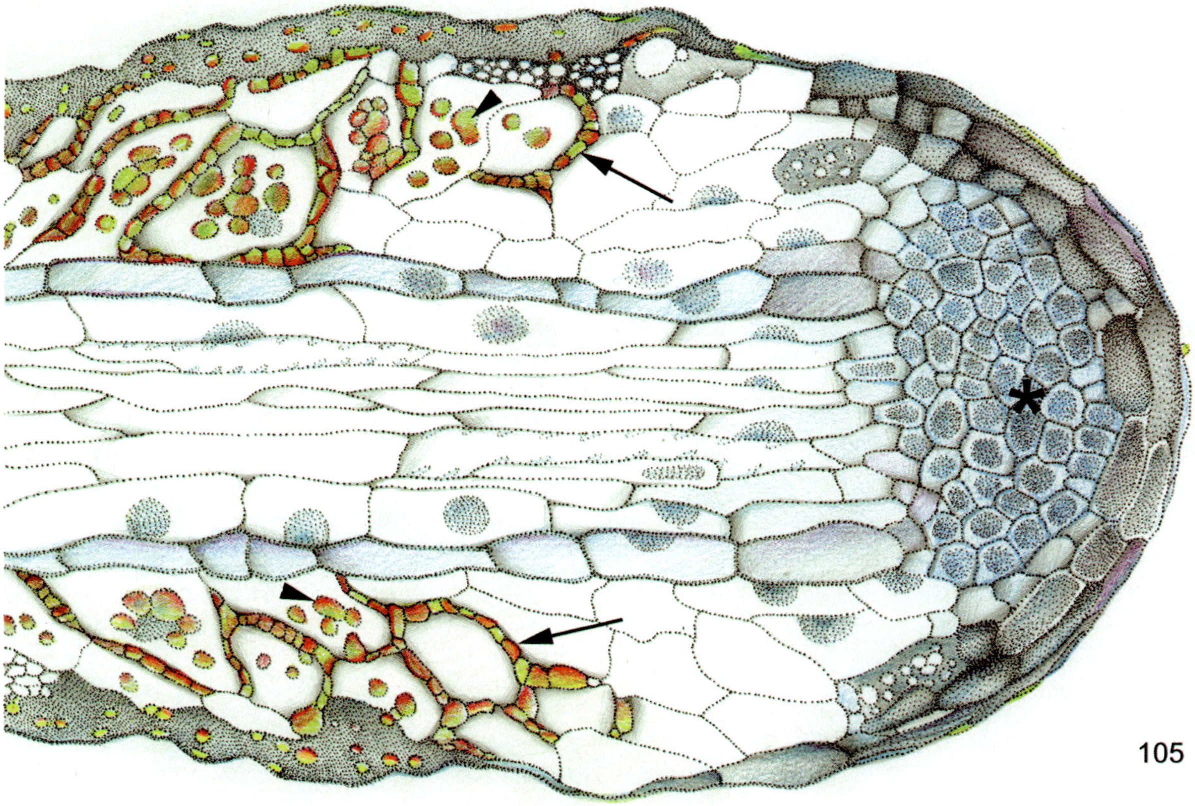

105

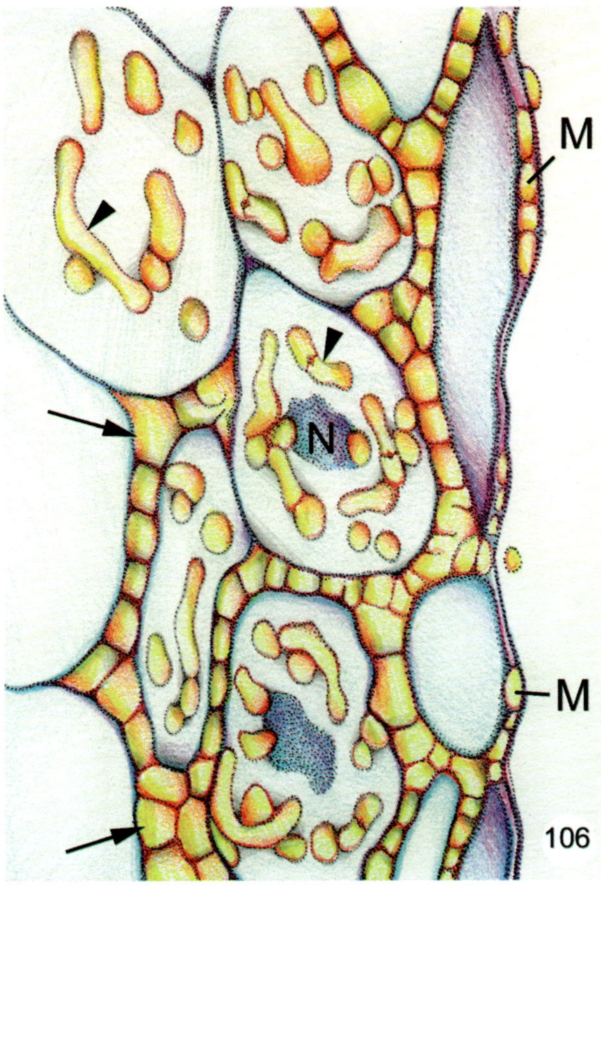

106

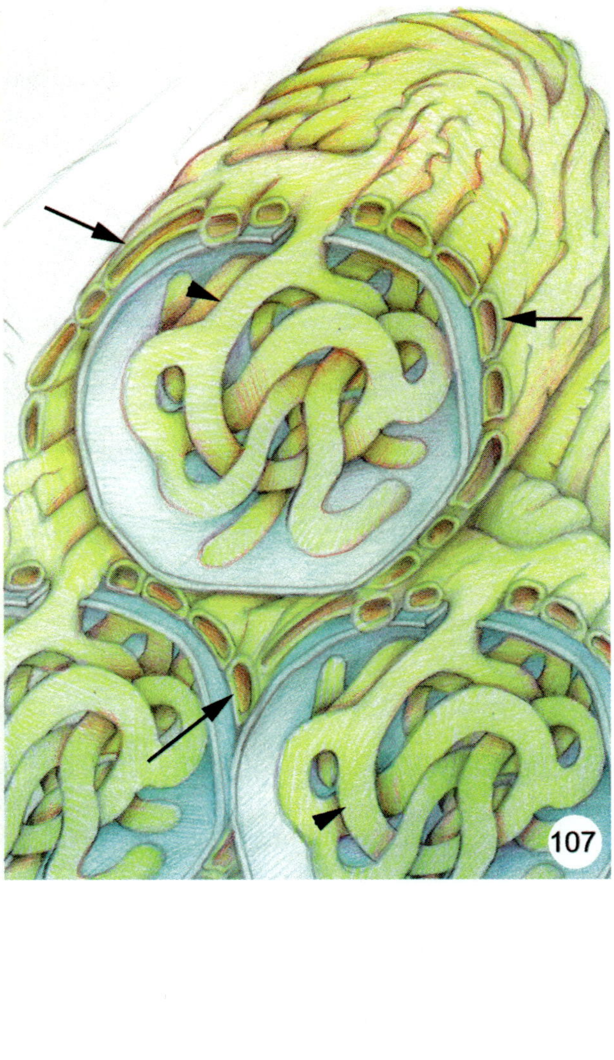

107

Figure 108. Hyphae of *Sphaerosporella brunnea* viewed with scanning electron microscopy. Ornamentations (arrowheads) are typical of this fungus. From Yu et al. Mycorrhiza, **11**: 167-177. (2001*a*)

Figure 109. Chlamydospore of *Wilcoxina mikolae*.

Figure 110. Apothecia of *Wilcoxina mikolae* var. *mikolae*. Photo courtesy of D. Luoma.

Figure 111. Dichotomous branching of root tip of *Pinus contorta* induced by colonization with *Sphaerosporella brunnea*.

Figure 112. Coralloid roots of *Pinus strobus* collected from a seedling nursery. Roots are colonized by an E-strain fungus.

Figure 113. *Larix* sp. roots colonized by an E-strain fungus.

Figure 114. *Pinus contorta* root colonized by an E-strain fungus showing features of the mantle.

Figure 115. Colonization of root hairs (arrowhead) of *Pinus banksiana* by hyphae of *Wilcoxina mikolae* var. *mikolae*.

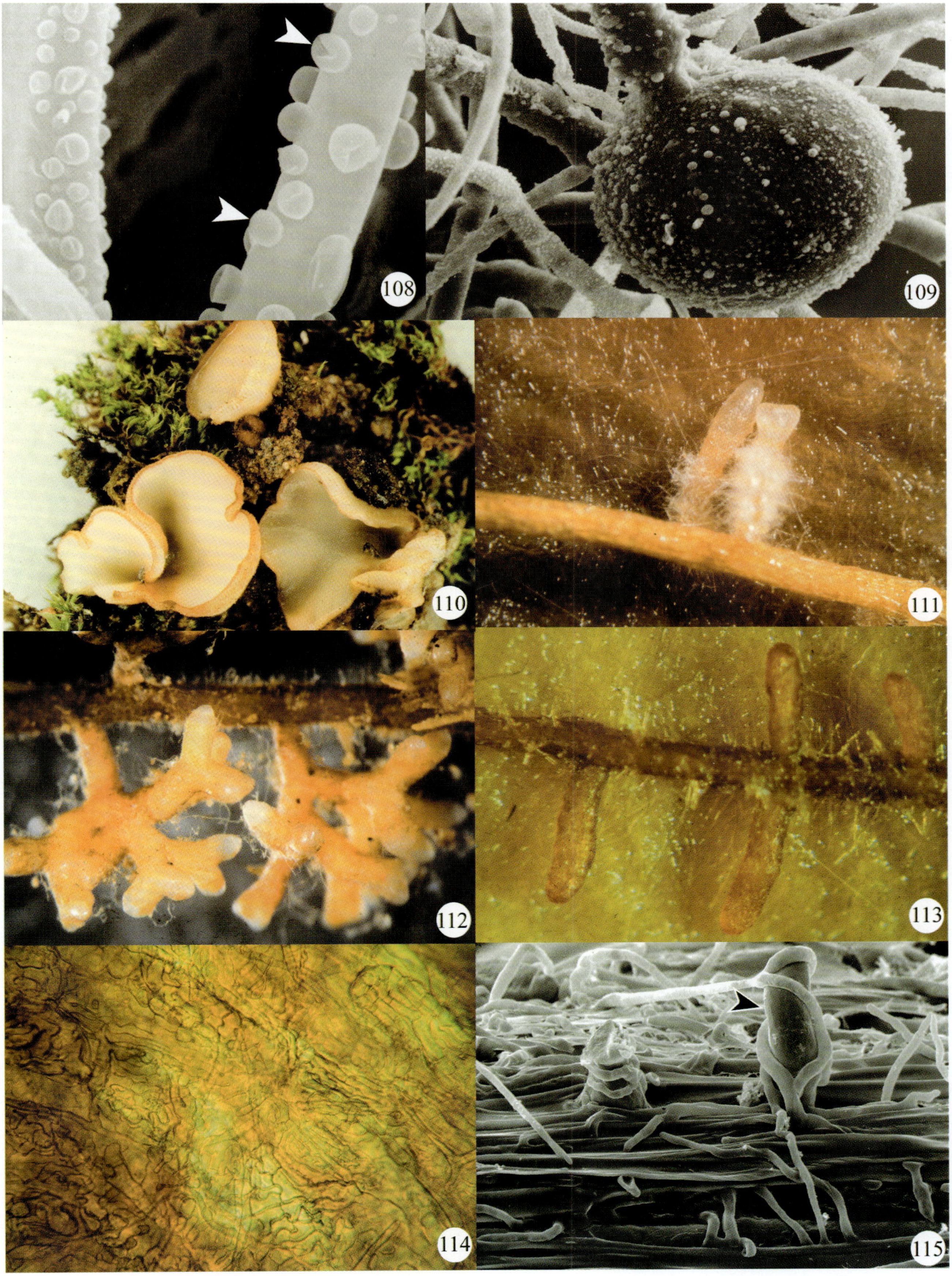
108
109
110
111
112
113
114
115

Figure 116. *Pinus resinosa* root showing attachment of hyphae of *Wilcoxina mikolae* var. *mikolae* to a root hair.

Figures 117–118. Mantle of *Pinus banksiana-Wilcoxina mikolae* var. *mikolae* ectendomycorrhizas showing hyphae (arrowheads) within mucilage (*) on the root surface.

Figures 119–120. Sections of *Pinus resinosa-Wilcoxina mikolae* var. *mikolae* ectendomycorrhizas showing the thin mantle (arrowhead), Hartig net hyphae (arrows), intracellular hyphae (double arrowheads), and host cell nuclei (n). From Yu et al. Mycorrhiza, **11**: 167-177. (2001*a*)

Figure 121. Transverse section of *Pinus banksiana-Wilcoxina mikolae* var. *mikolae* ectendomycorrhiza stained for polysaccharides. The mantle (m), Hartig net (arrows), and abundant intracellular hyphae (arrowheads) are evident.

Figure 122. Longitudinal section of dichotomous ectendomycorrhiza of *Pinus banksiana-Wilcoxina mikolae* var. *mikolae*.

Figure 123. *Pinus sylvestris* root. Intracellular hyphal complexes of an E-strain fungus are evident (arrowheads). The hyphal complex indicated by the black arrowhead has been removed from a cortical cell.

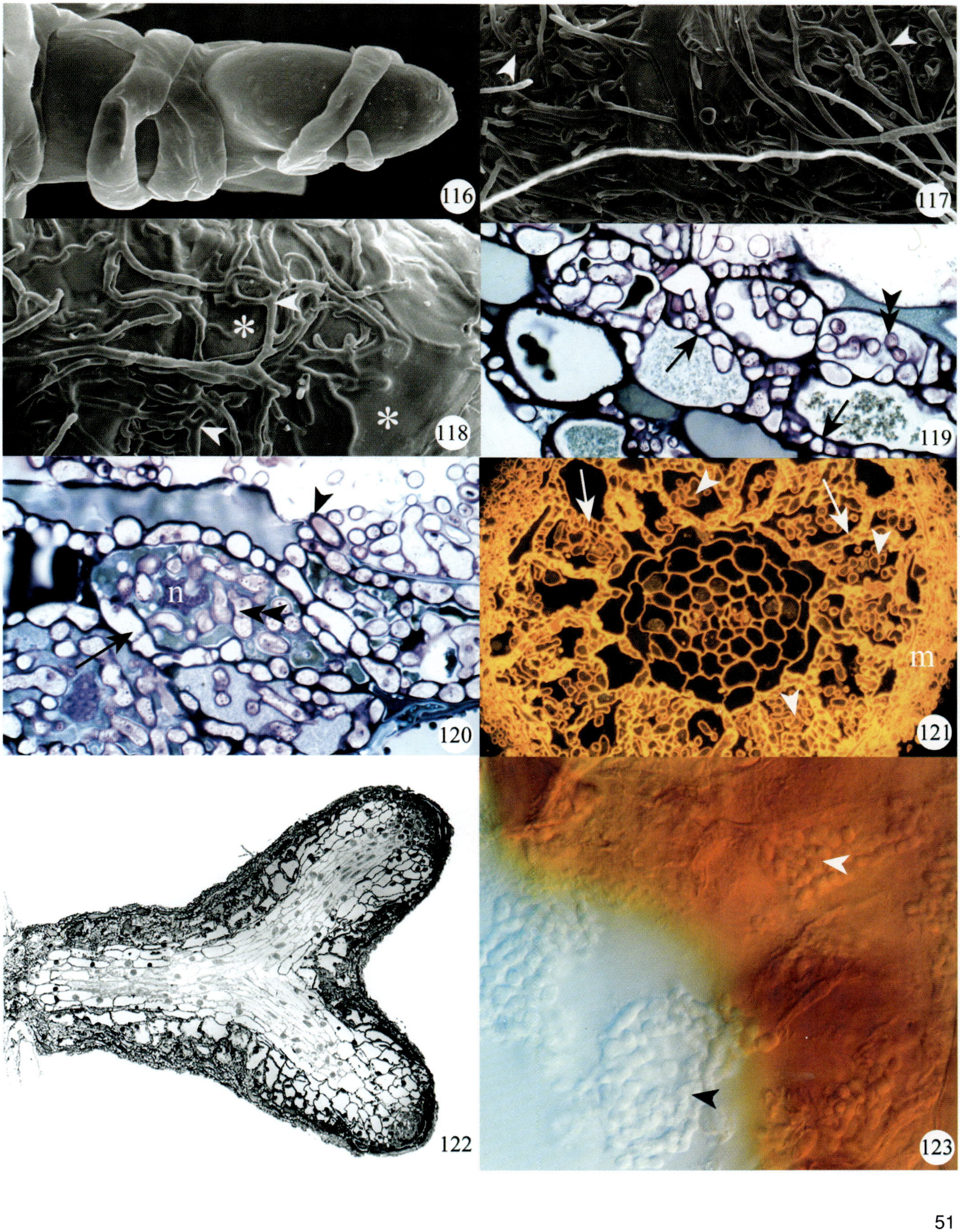
116
117
118
119
n
120
m
121
122
123

Figures 124–128. Electron microscopy of *Pinus banksiana-Wilcoxina mikolae* var. *mikolae* ectendomycorrhizas.

Figure 124. SEM showing branching of intracellular hypha (arrowheads).

Figures. 125–128. TEMs showing details of intracellular hyphae (Figure 125) and Hartig net hyphae (126–128).

Figure 125. Portion of an intracellular hypha showing mitochondria (arrowheads).

Figure 126. Hartig net hyphae with mitochondria (arrowheads).

Figure 127. Hartig net hyphae showing typical ascomycete Woronin bodies (arrowheads) along a septum (arrow).

Figure 128. Septa (arrows) and branching (arrowheads) of Hartig net hyphae. The dark inclusions within vacuoles are likely polyphosphate.

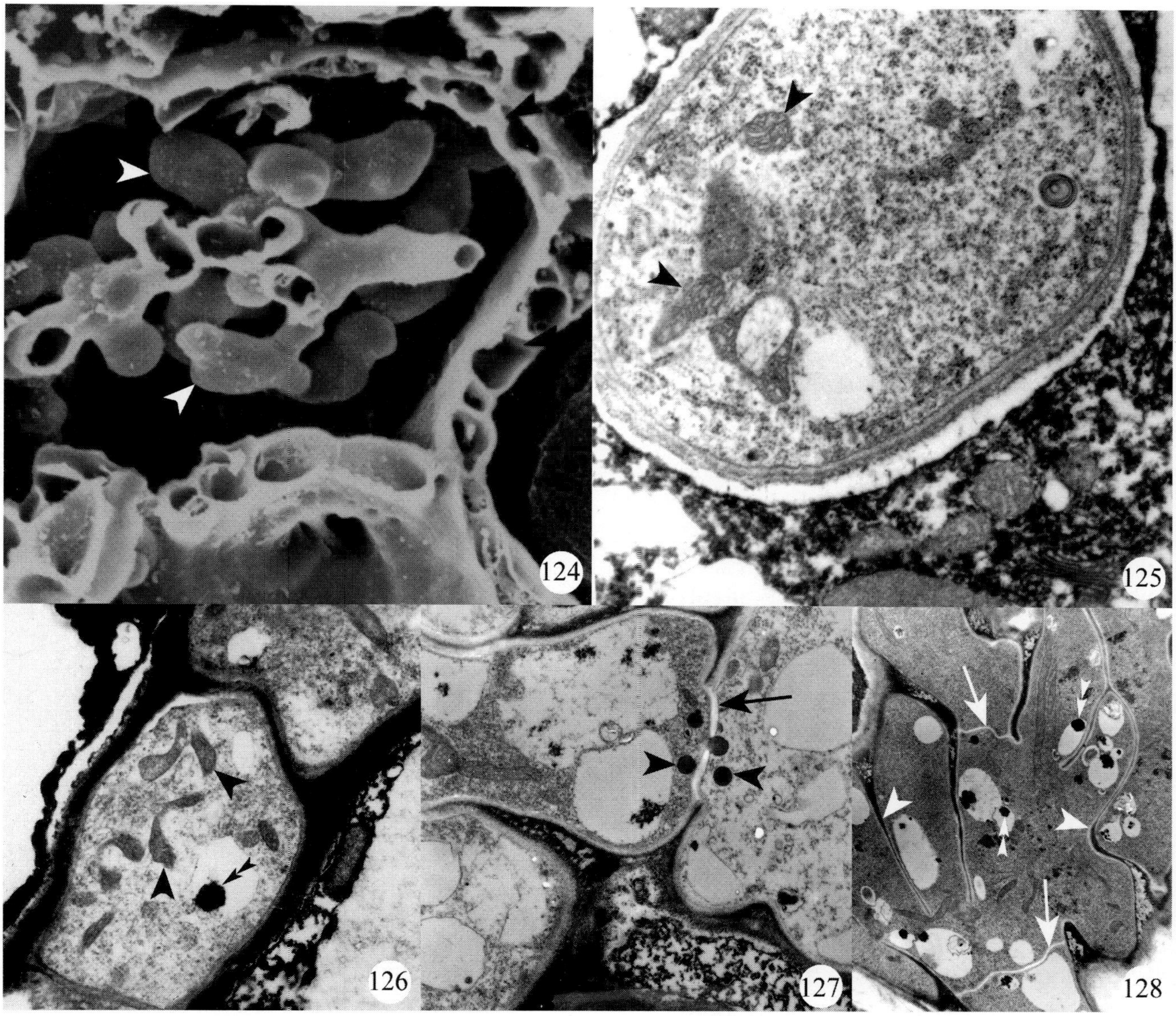
124
125
126
127
128

Intracellular hyphae develop a hyphal complex within epidermal and cortical cells that remains intact when fresh roots are squashed under a cover glass (Figure 123). The branching nature of the hyphal complex is evident in material examined with SEM (Figure 124). The host cell nucleus is usually surrounded by the hyphal complex (Figure 120). Intracellular hyphae are separated from the host cell cytoplasm by the elaboration of host plasma membrane (Figure 125); these hyphae and Hartig net hyphae have numerous mitochondria as well as small vacuoles with dense deposits, perhaps polyphosphate (Figures 125–128). Branched Hartig net hyphae (Figure 128) and intracellular hyphae frequently show Woronin bodies, typical of ascomycetous fungi (Figure 127). Although extraradical hyphae do develop, the extent of their development in the soil has not been assessed and there is no experimental evidence yet that they are involved in transport of nutrients to roots.

C. Functions

There is limited information concerning the contribution of the Hartig net and intracellular hyphae to the functioning of ectendomycorrhizas. To date, few definitive studies show benefits to plants that have ectendomycorrhizas. *Pinus* and *Larix* seedlings growing in disturbed sites frequently have ectendomycorrhizas and in some cases presumably benefit from the association. It is known that some fungal species forming ectendomycorrhizas are able to break down complex carbohydrates; perhaps sugars are being transferred to young seedlings before they develop full photosynthetic capacity. *Wilcoxina mikolae* and *Wilcoxina rehmii* are able to produce the siderophore, ferricrocin. Therefore, in mine spoils containing high concentrations of heavy metals such as iron, ectendomycorrhizas formed with *Wilcoxina* spp. may protect plants from toxicity induced by iron.

The high incidence of ectendomycorrhizas reported on *Pinus* and *Larix* species in seedling nurseries may be related to the cultural practices. In one study, a correlation was found between the incidence of ectendomycorrhizas and root rot symptoms on *Pinus strobus* seedlings, suggesting that ectendomycorrhizas may be indicative of poor seedling health (Ursic et al. 1997). A survey of urban white (*Picea glauca)* and blue or Colorado (*Picea pungens)* spruce showed a high occurrence of roots with ectendomycorrhizal associations, particularly on young trees (Danielson and Pruden 1989). The fungal species involved appeared to be tolerant to the dry, moderately alkaline soil conditions in the urban setting. Much work remains to be done with ectendomycorrhizas from a structural and ecological perspective. Ecologically, *Wilcoxina* species are likely able to form ectomycorrhizas with numerous hosts, and have the potential to link tree species in a fungal network.

Chapter 3. Arbuscular mycorrhizas

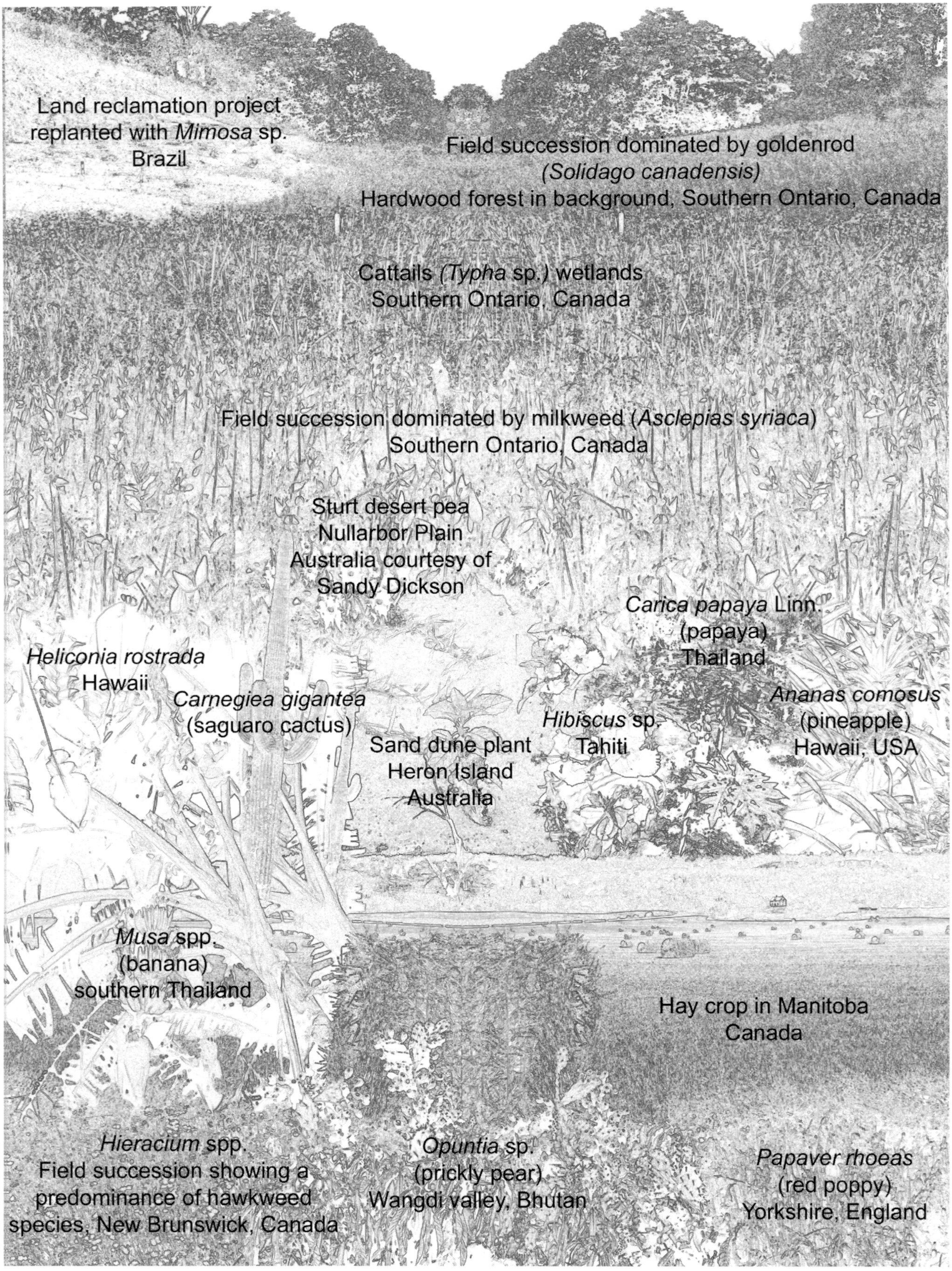
Land reclamation project replanted with *Mimosa* sp. Brazil
Field succession dominated by goldenrod (*Solidago canadensis*) Hardwood forest in background, Southern Ontario, Canada
Cattails (*Typha* sp.) wetlands Southern Ontario, Canada
Field succession dominated by milkweed (*Asclepias syriaca*) Southern Ontario, Canada
Sturt desert pea Nullarbor Plain Australia courtesy of Sandy Dickson
Carica papaya Linn. (papaya) Thailand
Heliconia rostrada Hawaii
Carnegiea gigantea (saguaro cactus)
Hibiscus sp. Tahiti
Ananas comosus (pineapple) Hawaii, USA
Sand dune plant Heron Island Australia
Musa spp. (banana) southern Thailand
Hay crop in Manitoba Canada
Hieracium spp. Field succession showing a predominance of hawkweed species, New Brunswick, Canada
Opuntia sp. (prickly pear) Wangdi valley, Bhutan
Papaver rhoeas (red poppy) Yorkshire, England

Chapter 3. Arbuscular mycorrhizas

A. Introduction

1. Definition

Arbuscular mycorrhizas, originally referred to as vesicular–arbuscular mycorrhizas, a name still used by some authors (Smith and Read 1997), are mutualistic symbiotic associations between the roots of most vascular plants and a small group of fungi in the new phylum Glomeromycota (Schüßler et al. 2001). Although some structural variation exists in this category, most arbuscular mycorrhizas are characterized by the presence of intraradical hyphae (intercellular or intracellular in location), arbuscules (finely branched hyphae involved in nutrient exchange), extraradical mycelium (hyphae that connect the root to the soil), and spores formed in the extraradical mycelium (Figures 129–131). Some fungal species also form intraradical structures referred to as vesicles (enlarged portions of hyphae that become filled with lipid bodies (Figures 130, 131), giving this group its original name vesicular–arbuscular mycorrhiza. Species in the genera *Gigaspora* and *Scutellospora* produce auxiliary vesicles (sometimes called auxiliary bodies or cells) in the extraradical mycelium (Figure 132).

Two major types of arbuscular mycorrhizas have been described: the *Arum*-type (Figure 130) and the *Paris*-type (Figure 131) named after the plant genera in the families Araceae and Liliaceae, respectively, in which they were first described (Gallaud 1905; Smith and Smith 1997). The differences between these types are elaborated on later in this chapter. Since some of the fungal structures of both types are formed within root cells, arbuscular mycorrhizas are classified under the broader term, endomycorrhizas.

Unlike ectomycorrhizas, arbuscular mycorrhizas do not produce an obvious change in the root system and the distinguishing characteristics can only be observed with the help of various microscopical methods. Using bright field or fluorescence microscopy, the fungal structures can be observed in fresh roots of some species, including onions (*Allium porrum*), strawberries (*Fragaria* sp.) and sorghum (*Sorghum sudanense*). However, roots usually must be processed in order to detect internal fungal structures. Typically, for routine observations of arbuscular mycorrhizas, roots are cleared of most of their cytoplasmic contents and then stained to view the fungal structures using light microscopy. Roots of a few host species synthesize a yellow pigment when colonized by arbuscular mycorrhizal fungi; however, this is inconsistent and cannot be used as a diagnostic feature for this association.

2. Plant species involved

It has been estimated that over 80% of all vascular plants form arbuscular mycorrhizas. A detailed list for the United Kingdom has been published (Harley and Harley 1987) but such lists generally do not exist for other regions of the world and, in fact, many species of vascular plants have not been assessed for the presence of mycorrhizas. Arbuscular mycorrhizas have been identified in a broad spectrum of plants, including some non-vascular plants, ferns and other seedless vascular plants **(see Box 6)**, groups within the gymnosperms including conifers (e.g., *Thuja, Sequoia, Metasequoia*), *Ginkgo biloba*, the cycads, and the majority of angiosperm families.

The few angiosperm families that do not have arbuscular mycorrhizas either form other categories of mycorrhizas or lack mycorrhizas. Among the latter families are the Brassicaceae (this family includes canola, mustards, cabbages, etc.) and the Chenopodiaceae (this family includes garden and sugar beets, spinach and the large genus *Chenopodium*), although even here arbuscular mycorrhizal associations have been reported for a few species. A few aquatic plant families and the sedges may have low levels of colonization by arbuscular mycorrhizal fungi, however, these can be overlooked if roots are collected at the wrong time of the year or if the sample size is too small. The relative abundance and the seasonality of arbuscular mycorrhizas in many plant species remains undetermined.

Fossil records from the Devonian of the first vascular plants show that arbuscular mycorrhizal associations were present in underground rhizomatous structures, confirming the presence of this mutualistic association as far back as 400 million years or more (Taylor et al. 1995; Figures 133, 134). Evidence that the first land plants already possessed this association suggests that their aquatic ancestors also harboured these fungi. Redecker et al. (2000) have documented spores from the Ordovician period (460 Ma) similar to present day spores of Glomalean fungi, indicating that earlier associations with non-vascular plants were likely.

Arbuscular mycorrhizas have been studied intensively; they occur in most ecosystems of the world and are found in many important crop species (e.g., wheat, maize, rice, soybeans, forage crops, grapes, fruit trees, cotton) and horticultural species (e.g., roses, petunias, lilies, carnations). Attempts have been made to increase productivity by adding arbuscular mycorrhiza inoculum during planting or by manipulation of soil conditions to enhance the performance of indigenous fungal species. The inability to axenically culture arbuscular mycorrhizal fungi has made it difficult to produce inoculum on an industrial scale and consequently has limited its use in large agricultural operations.

3. Fungal species involved

The fungi involved are ubiquitous soil-borne organisms (Phylum Glomeromycota; Class Glomeromycetes) belonging to four orders: Archaeosporales, Paraglomerales, Diversisporales and Glomerales (Schüßler et al. 2001). Eight genera of arbuscular mycorrhizal fungi have been recognized based mainly on morphological characteristics of asexual spores, although molecular methods and various biochemical parameters are now being used in systematic studies. These genera, *Glomus, Paraglomus, Sclerocystis, Scutellospora, Gigaspora, Acaulospora, Archaeospora,* and *Entrophospora,* include approximately 150 species; however, species delineation remains uncertain and continues to change as more isolates are examined and as the use of molecular techniques increases (Morton and Redecker 2001). A major challenge in the systematics of this group of fungi is the considerable variation within isolates of defined species and the lack of sexual reproductive structures.

All arbuscular mycorrhizal fungal species are obligate biotrophs, depending entirely on host plants for carbon compounds. This means that, unlike many ectomycorrhizal fungi **(see Box 5)**, these fungi cannot be cultured in the absence of plants. As a result, researchers have used root organ cultures and genetically transformed root cultures to maintain arbuscular mycorrhizal fungi and for use in experimental studies (Fortin et al. 2002) **(see Box 7)**.

B. Intraradical hyphae

1. Development

The formation of arbuscular mycorrhizas involves a series of steps from the recognition of the root surface by the fungus to the formation of an appressorium, epidermal cell penetration, intraradical hyphal and arbuscular development, and, in some cases, the formation of vesicles. All of these steps are undoubtedly under genetic control, but until recently, there has been little research in this area. Plant mutants that show blocks to various steps in the colonization process are now being used in such studies **(see Box 8)**.

In order for the development of all internal structures to occur, fungal hyphae must contact either the surface of root epidermal cells (Figures 130, 131, 135) or root hairs (Figure 136), form specialized attachment structures called appressoria (Figure 135) and then penetrate the epidermal layer before reaching the cortex. The regions along the root at which appressoria form and where hyphae enter the epidermis are referred to as entry points. An hypha, on contacting the root surface, may branch and form more than one entry point. Sources of hyphae that attach to the surface of epidermal cells or root hairs are either germinating spores, the existing hyphal network in the soil attached to living roots, or hyphae that grow from colonized root pieces left in the soil as plants die. Sectional views of appressoria show that they are multi-nucleate and possess many small vacuoles (Figure 137). Often an appressorium forms between epidermal cells (Figure 137) and hyphae formed from the appressorium penetrate adjacent epidermal cells (Figures 138, 139).

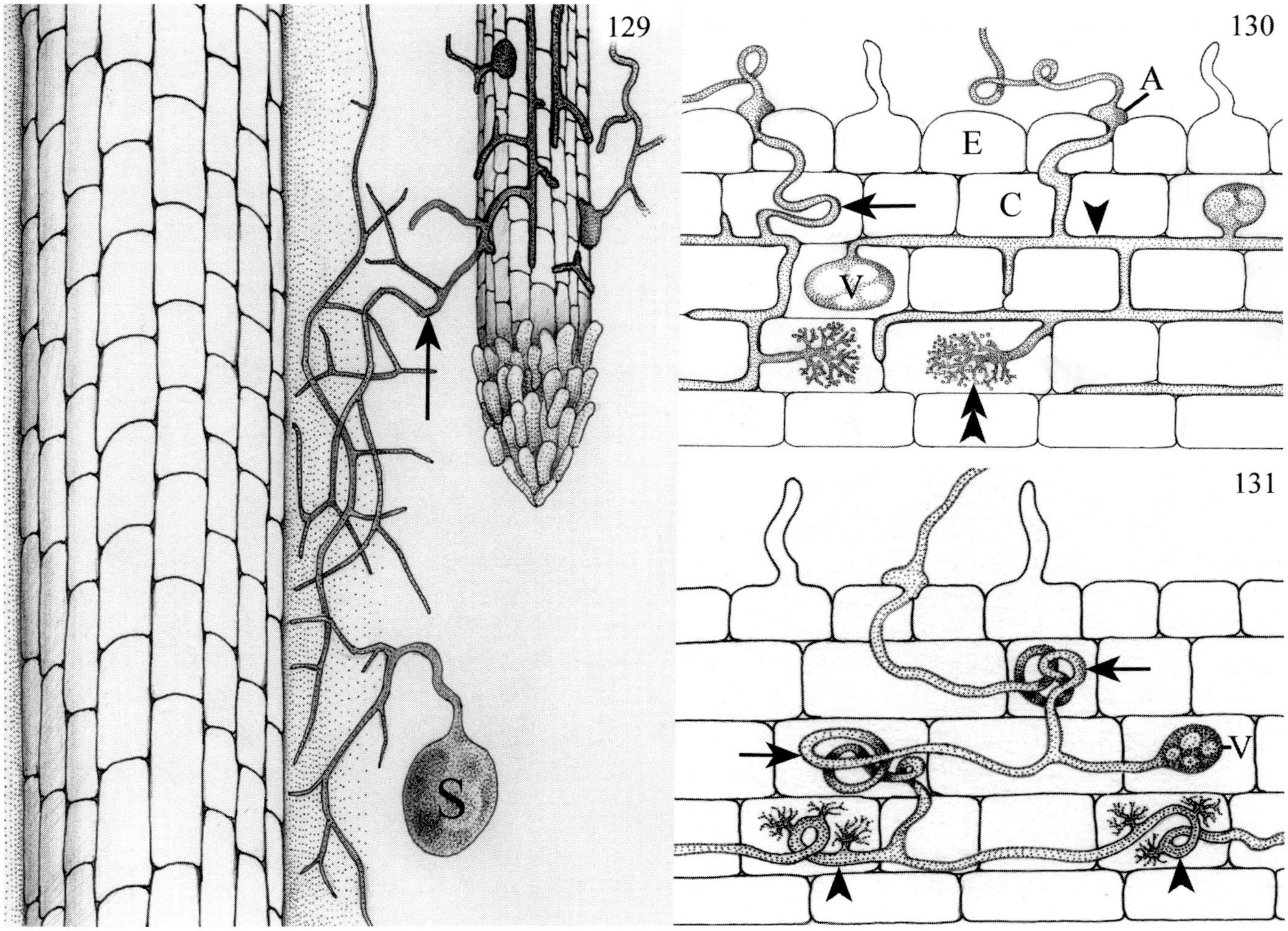

Figures 129–131. Diagrammatic representation of root colonization by arbuscular mycorrhizal fungi.

Figure 129. Colonization of root surface by hyphae either from a germinating spore (S) or from a previously colonized root (arrow).

Figure 130. *Arum*-type arbuscular mycorrhizal association. Infection hyphae from appressoria (A) penetrate epidermal cells (E) before entering the cortex (C). Frequently, a hypha forms a coil (arrow) before entering the intercellular space system (arrowhead) of the cortex. Arbuscules (double arrowheads) form within inner cortical cells and, depending on the fungal species, vesicles (V) may form.

Figure 131. *Paris*-type arbuscular mycorrhizal association. Early events are similar to the *Arum*-type but extensive hyphal coiling occurs in the cortex (arrows) and small branches form from some of these to form arbusculate coils (arrowheads). Again, depending on the fungal species, vesicles (V) may form.

Figure 132. Auxiliary vesicles of *Gigaspora* sp. formed in the extraradical mycelium. Photo courtesy of Dr. Mark Brundrett.

Figures 133, 134. Fossil arbuscular mycorrhizal material from the Early Devonian. An hypha (arrowhead) with an attached vesicle (V), and an arbuscule (double arrowheads) are evident. From Taylor et al. Mycologia **87:** 560–573 (1995).

Figure 135. An appressorium (arrowhead) and branching hyphae (double arrowhead) of *Glomus versiforme* on the surface of an *Allium porrum* root.

Figure 136. Entry of hypha into a root hair of a *Panax quinquefolius* (ginseng) root.

Figures 137–138. Stages in the penetration of epidermal cells of *Allium porrum* by *Glomus versiforme* hyphae.

Figure 137. Sectional view of an appressorium. Nuclei (arrowheads) and small vacuoles (double arrowheads) are evident. Figure 137 from Garriok et al. *New Phytol*. **112**: 85-92. (1989).

Figure 138. Early stage in the formation of penetrating hyphae (arrowheads).

132

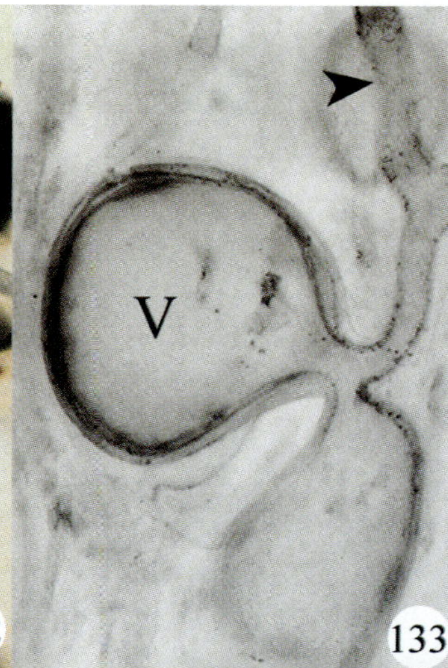
V
133

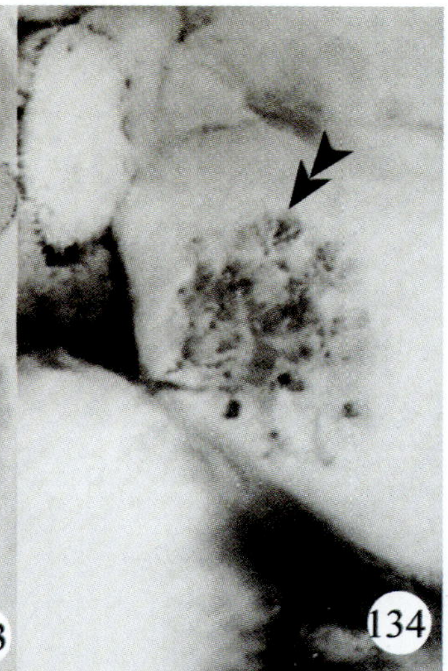
134

135

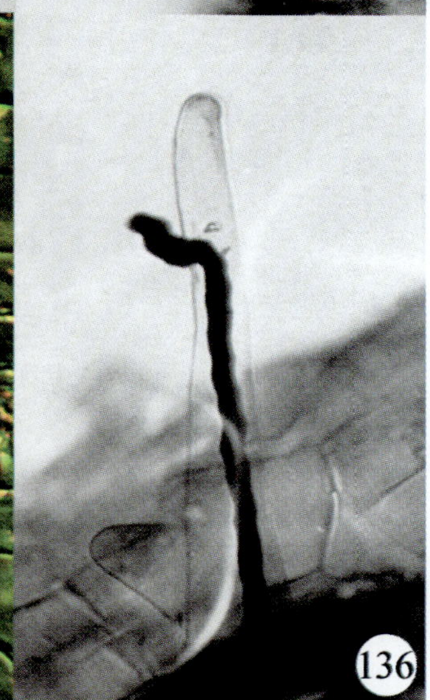
136

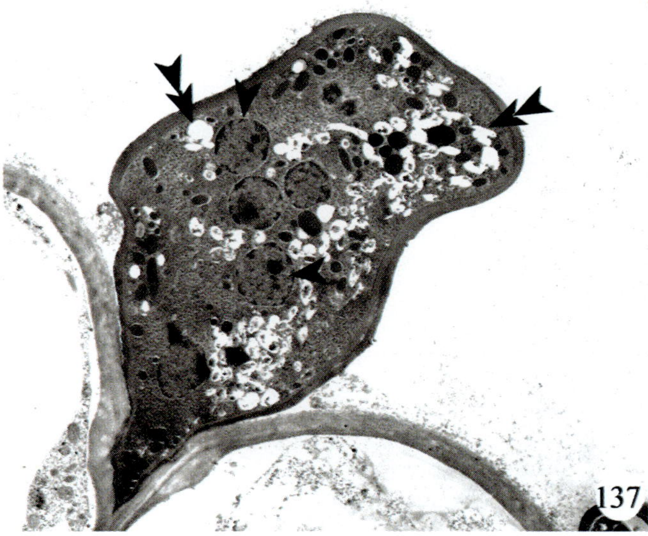
137

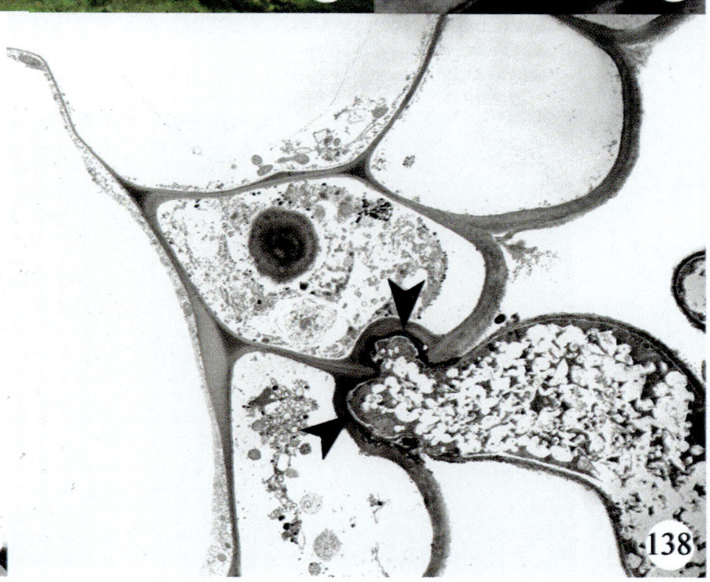
138

Box 6: Do the 'lower' plants have arbuscular mycorrhizas?

With respect to research on arbuscular mycorrhizas (AM) in angiosperms and gymnosperms, relatively little research has been done on plant groups such as the liverworts, hornworts, mosses, lycopods and ferns. Most studies are surveys of species to determine the presence or absence of AM associations but few include experimental work on function (see Read et al. 2000 for a review). One of the more detailed surveys is a paper on ferns and fern allies in Yunnan, southwest China, in which 256 species were examined for arbuscular mycorrhizas (Zhi-wei 2000). A limitation with surveys of this type is that the sample size for any one species is often low and collections are frequently made at only one time of the year. This may account for reports of absence of arbuscular mycorrhizas in some plants. For example, Zhi-wei (2000) found that the majority of fern allies and the higher ferns (leptosporangiate ferns) were non-mycorrhizal whereas all of the lower ferns (eusporangiate ferns) were mycorrhizal. Although he argues that results show an evolutionary trend to loss of mycorrhizas in these groups, additional studies are required to clarify these relationships.

The achlorophyllous gametophytes of the eusporangiate ferns and members of the lycopods are all mycorrhizal. The morphology of the association most resembles the *Paris*-type arbuscular mycorrhiza; it displays coiled hyphae, few if any arbuscules, and presence or absence of vesicles.The fungal endophytes in gametophytes of *Lycopodium clavatum* consist of fine, coiled, aseptate hyphae and small vesicles (Schmid and Oberwinkler 1993). Green, epigeous gametophytes as well as roots of species in the fern family Gleicheniaceae, have typical arbuscular mycorrhizas in that intracellular hyphae, arbuscules and vesicles form (Schmid and Oberwinkler 1995). Other 'lower' plant species remain to be studied.

Mosses, a group of 'lower' plants, consistently lack associations with symbiotic fungi (Read et al. 2000). It is not known why this group resists mycorrhizal fungi; however, experiments could be designed to explore possible mechanisms involved.

Some liverwort species associate with fungi normally found with ericoid plants (Duckett and Read 1995), others with arbuscular mycorrhizal fungi (Read et al. 2000) and still others with basidiomycetes that normally form ectomycorrhizas with tree species (Read et al. 2000). Experimentally, Read et al. (2000) have shown that thalli of the liverwort, *Pellia*, when grown in proximity with *Plantago* plants, become colonized with arbuscular mycorrhizal fungi; thallus cells contain hyphae, arbuscules and vesicles. Other experiments with liverworts are discussed in the chapters on ectomycorrhizas and ericoid mycorrhizas.

Figure a. *Osmunda* and other fern species. **Figure b**. Croziers (fiddleheads), typical of most ferns.

Box 7: Use of "hairy roots" in studies with arbuscular mycorrhizal fungi

"Hairy root" is a pathogenic condition induced by the transfer of root-inducing (Ri) plasmid T-DNA from the bacterium, *Agrobacterium rhizogenes,* to various plant species. This plasmid T-DNA can be transferred into hypocotyls, epicotyls or radicles of sterile seedlings and the resulting induced adventitious or lateral roots occurring at the infection site can be excised and grown in axenic cultures. The morphological manifestation of the plasmid T-DNA transfer is the production of numerous adventitious and lateral roots and hence the term "hairy roots". Since the T-DNA is integrated into the plant genome, the transformation is stable and new cultures can be established by the excision and transfer of portions of the root system under axenic conditions (Bais et al. 2001). The usefulness of this system for studying the establishment and physiology of arbuscular mycorrhizas was first recognized by Mugnier and Mosse (1987), but now "hairy root" cultures of a number of host species combined with several arbuscular fungal species are being used by many researchers (see review by Fortin et al. 2002).

Because these cultures are axenic and free of soil, they are especially useful in tracking the early events in spore germination and chemotropic response of germ tubes to plant roots (Bécard and Fortin 1988) and the development and branching patterns of extraradical mycelium (Bago et al. 1998*a,b*), events that are difficult to study in soil. "Hairy root" cultures have also been used to study the influence of roots (Bécard and Piché 1989*a*) and factors such as CO_2 (Bécard and Piché 1989*b*) and flavonoids (Bel Rhlid et al. 1993) on fungal growth. They have also been used to assess the interaction between root pathogenic fungi and other soil organisms and arbuscular mycorrhizal fungi (Benhamou et al. 1994; Fortin et al. 2002) as well as the interaction between arbuscular mycorrhizal fungi and nodulation mutants of *Pisum sativum* (Balaji et al. 1994; 1995) and *Medicago truncatula* (Boisson-Dernier et al. 2001).

"Hairy root" cultures are amenable to experiments involving nutrient uptake and to those assessing the effect of the extraradical mycelium on substrate pH (Bago et al. 1996; Villegas et al. 1996). "Hairy root" and other root-organ cultures are now being used routinely to produce mycelium and spores of various arbuscular mycorrhizal species for systematic and phylogenetic studies (Fortin et al. 2002).

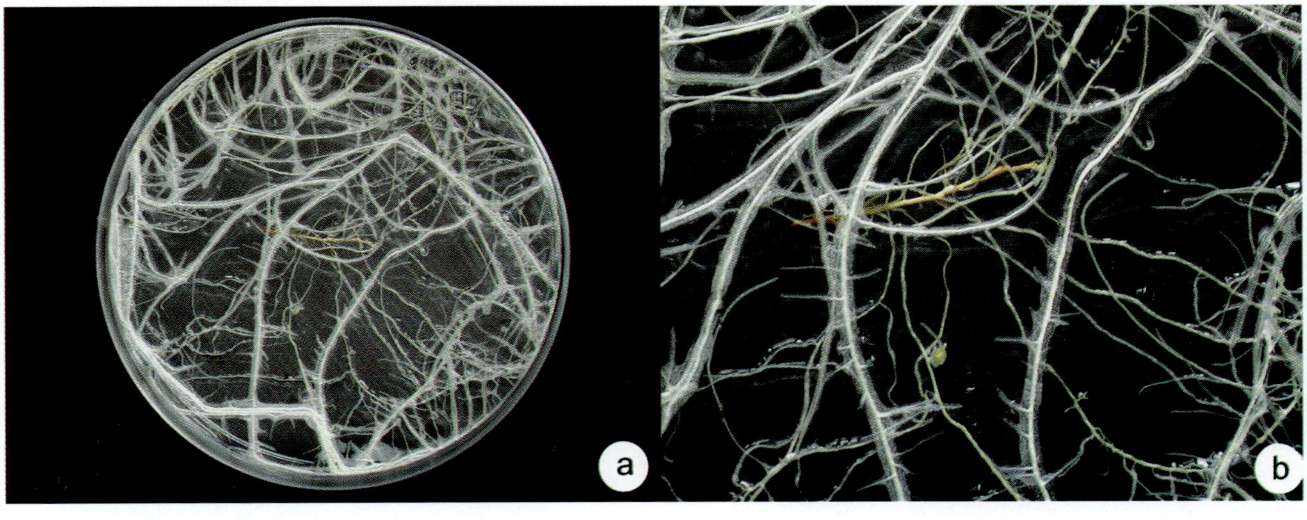

Figures a and b. Transformed *Daucus carota* roots.

In the *Arum*-type arbuscular mycorrhiza, the hypha that penetrates into the epidermis generally forms a coil either in the epidermal cell or first cortical cell layer (Figures 130, 141) before it enters the intercellular spaces of the cortex (Figures 130, 142, 143). In this type, roots can become rapidly colonized along the root axis due to the free movement of hyphae in the intercellular spaces.

In the *Paris*-type mycorrhiza, the intraradical hyphae pass from cell to cell, forming complex coils in both epidermal and cortical cells (Figure 131). Species of plants developing *Paris*-type mycorrhizas lack conspicuous intercellular spaces in the root cortex so that the growth of hyphae in the longitudinal direction of the root is slower than in the *Arum*-type.

Many angiosperm roots develop a modified outer cortical cell layer, the hypodermis (known as an exodermis if it has Casparian bands, wall modifications consisting primarily of the hydrophobic substance, suberin), which has cells that differ in their length in reference to the longitudinal axis of the root. This type of hypodermis or exodermis is referred to as dimorphic, consisting of short and long cells (Figure 140). Both cell types become modified by the deposition of suberin in their walls, a process that is slower in the short cells compared to the long cells. The short cells, therefore, become the entry points for arbuscular mycorrhizal fungal hyphae into the remaining cortex (Figure 141).

2. Functions

Intraradical hyphae are able to initiate other fungal structures within the root. For example, in the *Arum*-type mycorrhiza, hyphae penetrate through the walls of cortical cells, branch repeatedly, and form arbuscules (Figure 130). In the *Paris*-type, small branches develop laterally from the hyphal coils to form arbuscules (referred to as arbusculate coils; Figure 131). Intraradical hyphae may also initiate intracellular or intercellular vesicles (Figures 130, 131).

Intraradical hyphae that persist in decaying root pieces in the soil are important sources of inoculum for subsequent colonization of new roots. By withstanding soil freezing and drying, these hyphae can survive in the soil until conditions are favourable for their re-growth.

C. Arbuscules

1. Development

Arum-type

Usually a single branch from either an intercellular or intracellular hypha narrows and penetrates the wall of a cortical cell. Here it forms a trunk hypha which branches repeatedly to form a complex tree-like structure, the arbuscule (Figures 143, 144, 145). Occasionally two trunk hyphae form, leading to the development of two arbuscules within one cortical cell. Arbuscules may form in any region of the cortex but most often they develop in the inner cortex adjacent to the endodermis and the vascular tissues. The trunk hypha, and branches from it, become enveloped by host-derived plasma membrane (the periarbuscular/perifungal membrane) and a host-derived interfacial matrix consisting of cell wall components **(see Box 9)**. This results in the separation of the arbuscule from the host cell cytoplasm by an apoplastic compartment which must play a role in the transfer and temporary storage of mineral nutrients and sugars.

Paris-type

Plant species that have this type of mycorrhiza lack conspicuous intercellular spaces in their roots and, as a result, only intracellular hyphae are present. These develop complex coils (Figures 146, 147) from which fine lateral branches are initiated (Figure 148). The coils, including these fine branches, are the arbuscules or arbusculate coils. The hyphal coils and the fine branches have a perifungal membrane and an interfacial matrix of host-derived cell wall components **(see Box 9)**.

Appressoria, hyphae (both extraradical and intraradical), and arbuscules may all have endosymbiotic bacteria within the cytoplasm **(see Box 10)**.

The development of intracellular hyphae and arbuscules leads to dramatic changes in the organization of the cytoplasm in host cells. Cortical cells that were vacuolated prior to fungal penetration have an increased number of organelles such as mitochondria. One of the most striking changes, however, is in the cytoskeleton **(see Box 11)**.

Box 8: Plant mutants help unravel colonization events in arbuscular mycorrhizas

Plants with a mutation in one or more genes have been used extensively to determine the genetic control of developmental events. Duc et al. (1989) reported that mutants in some legume species, when inoculated with *Rhizobium,* either failed to form nodules (*nod*$^-$) or formed nodules that failed to fix atmospheric nitrogen (*nod*$^+$/*fix*$^-$). In addition, these were unable to form typical arbuscular mycorrhizal (AM) associations i.e., were *myc*$^-$. This led to a search for other plant mutants showing similar responses (Peterson and Bradbury 1995; Peterson and Guinel 2000). In several legume species, a number of mutants defective in nodulation have been identified and these also show blocks at various stages of the AM colonization process. This has led to the discovery of a number of common genes in the establishment of these two diverse symbioses in legumes (Hirsch and Kapulnik 1998; Harrison 1999).

Several legume mutants block the ingress of the fungus at the epidermis and these hosts are particularly useful in studying the initial colonization process since AM fungal hyphae must receive plant signals for attachment and appressorium formation to occur, a requirement for root penetration. In all *myc*-legume mutants described to date, signals for attachment and appressorium formation are produced and the block occurs subsequently. In some of these mutants, appressoria of abnormal structure form on the root surface; in others, there is an increase in the number of these structures. In *Lotus japonicus*, a model species for the analysis of the genetic/molecular control of legume–*Rhizobium* interactions, *nod*$^-$ mutants have been recovered after treatment with the chemical mutagen, ethylmethane sulfonate (EMS) that show a block of AM fungi either in the epidermis or the exodermis, the first layer of cortical cells (Bonfante et al. 2000; Senoo et al. 2000). In one of these mutants (*Ljsym4–2*), Bonfante et al. (2000) showed that hyphae of the AM fungus *Gigaspora margarita* formed appressoria and penetration hyphae that entered epidermal cells. However, epidermal cell death occurred, limiting hyphae to this outer layer.

In other mutants, interference in the colonization process occurs later, either blocking the colonization of the cortex or leading to the formation of abnormal arbuscules that develop only a few stumpy branches (Gianinazzi-Pearson 1996). A mutant with reduced ability to form AM associations has been identified in a non-legume species, *Lycopersicon esculentum* (Barker et al. 1998). The colonization patterns of this mutant are dependent on the AM fungal species used as the inoculum (Gao et al. 2001).

There is tremendous scope for future studies with plant mutants in determining the molecular basis of the complex colonization process in arbuscular mycorrhizas.

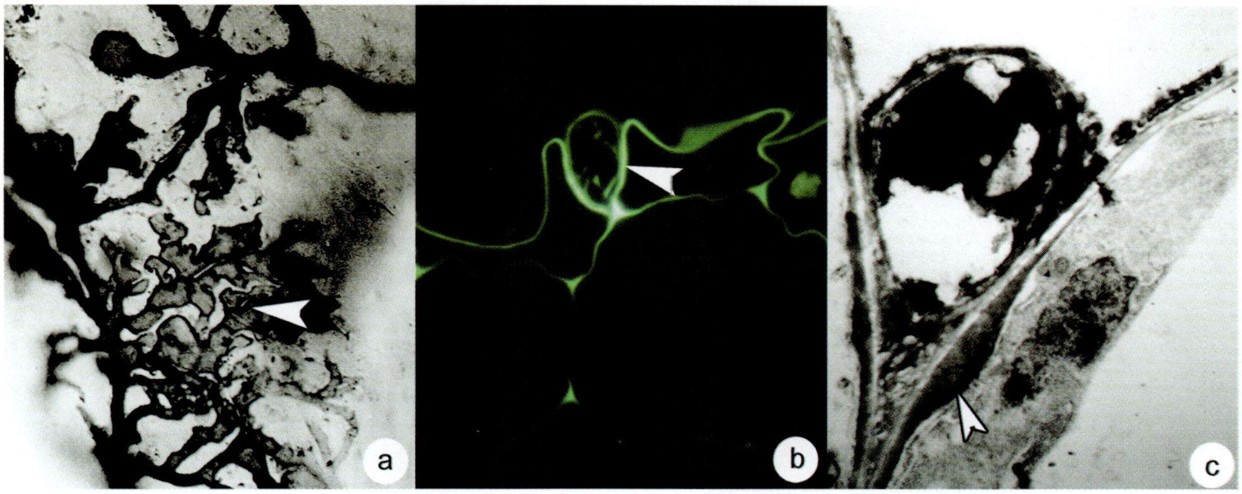

Figure a. Abnormal appressoria (arrowheads) of *Glomus fasciculatum* on root surface of a *nod*$^-$ *fix*$^-$ genotype of *Medicago sativa*. **Figures b and c**. Appressoria of *Glomus versiforme* on root surface of a *nod*$^-$ *fix*$^-$ genotype of *Medicago sativa*. Figure b. Epidermal wall thickening (arrowhead) occurs adjacent to the appressorium. Figure c. Appressorium with degenerated cytoplasm. Wall thickening (arrowhead) is present.

Figure 139. Penetrating hypha (arrowheads) of *Glomus versiforme* has entered an epidermal cell of an *Allium porrum* root.

Figure 140. Root of *Asparagus officinalis* with a dimorphic exodermis. Short cells (arrowheads) alternate with long cells (*).

Figure 141. Hyphal coil of *Glomus intraradices* within a short exodermal cell of an *Asparagus officinalis* root.

Figure 142. An intercellular hypha of *Glomus versiforme* in an *Allium porrum* root.

Figure 143. Intercellular hypha (arrowheads) and arbuscules (arrows) of *Glomus mosseae* in a cleared root of *Allium porrum*.

Figure 144. Early stage of arbuscule formation of *Glomus mosseae* in cleared root of *Allium porrum*.

Figure 145. Fully developed arbuscule of *Glomus mosseae* in cortical cell of an *Allium porrum* root.

Figure 146. Hyphal coils of *Glomus intraradices* in cortical cells of a *Panax quinquefolius* root.

Photos in Figures 143–145 provided by Dr. M. Brundrett.

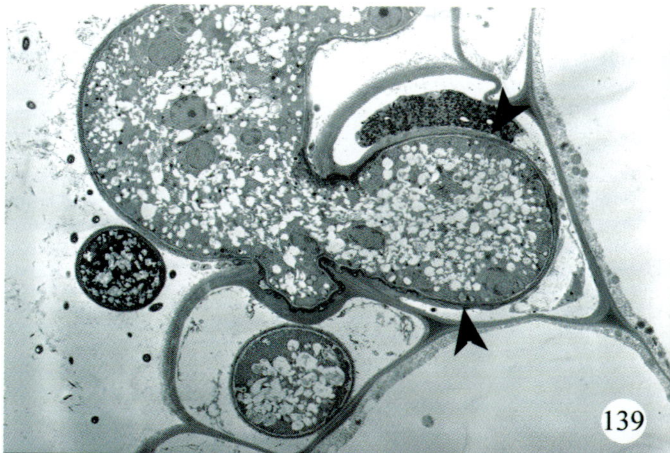

139

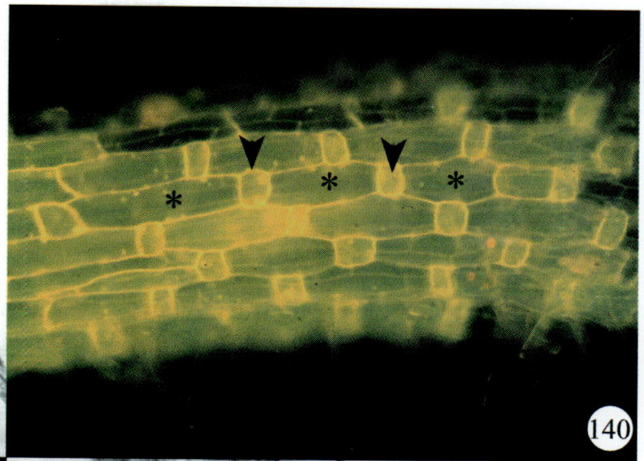

140

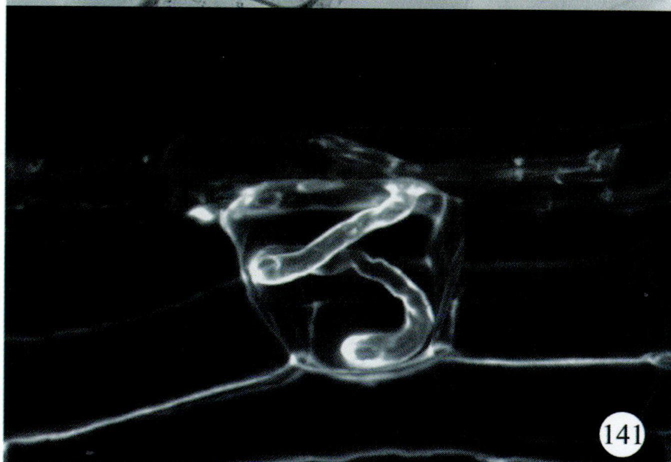

141

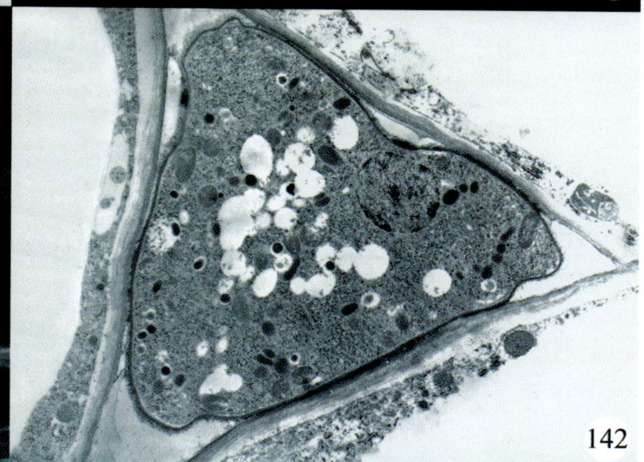

142

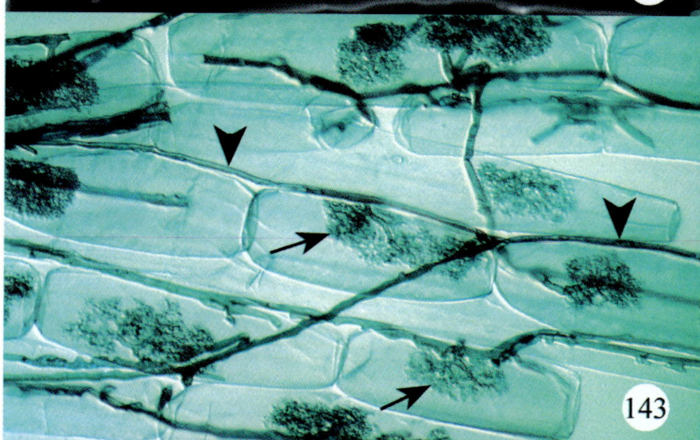

143

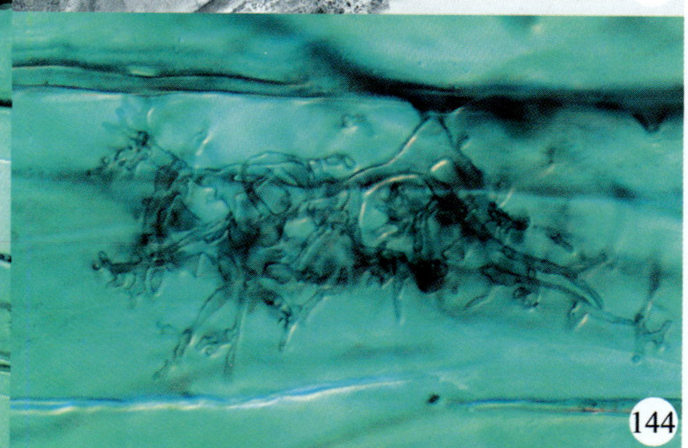

144

145

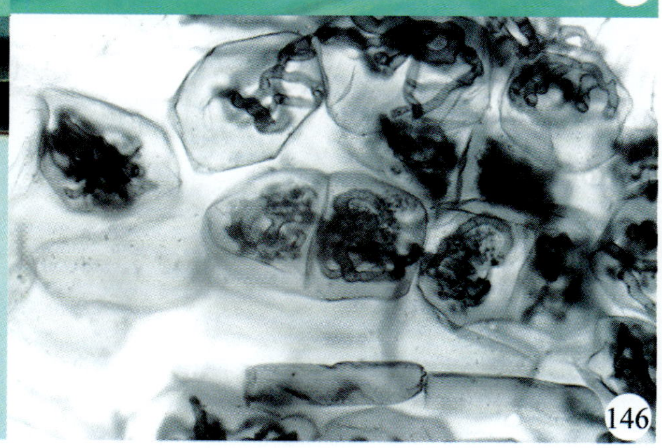

146

Box 9: What is the nature of the interface between arbuscular mycorrhizal fungi and root cells?

In both *Arum*-type and *Paris*-type arbuscular mycorrhizas, all intracellular structures (hyphae, arbuscules and vesicles, if present) are separated from the root cytoplasm by a host-derived membrane and an interfacial matrix, effectively restricting the fungus to the apoplast. Research, until recently, has focused on the interface formed between arbuscules and root cell cytoplasm in the *Arum*-type association (Bonfante and Perotto 1995). This interface is presumed to be the main site of carbon uptake from root cells by the fungus, and for the release of mineral nutrients into root cells by the fungus. The host plasma membrane is elaborated over all branches of the arbuscule to form the periarbuscular membrane. This membrane increases the surface area presumably for bi-directional transport. It differs from the plasma membrane adjacent to the cell wall in that it has high ATPase activity (Smith and Smith 1990), supporting the hypothesis that it may be involved in active transport of nutrients.

An interfacial matrix or compartment exists between the fungal cell wall and the periarbuscular membrane in the *Arum*-type association. Studies by Bonfante and co-workers (review in Bonfante and Perotto 1995) using various affinity methods combined with transmission electron microscopy, have shown that this region consists of host-derived cell wall components, including β–1,4-glucans, pectins, xyloglucans, aribinogalactans, rhamnogalacturonans, hydroxy-rich glycoproteins, and non-esterified homogalacturonans. The formation of this new apoplastic compartment is a main feature of this biotrophic association (Bonfante and Perotto 1995) but its functional role is still not clear.

In the *Paris*-type arbuscular mycorrhiza, Armstrong and Peterson (2002) have shown that both the hyphal coils and arbusculate coils are surrounded by membrane and interfacial matrix material. The fine branches of arbusculate coils have less matrix material than the hyphal coils but the fact that both are surrounded by membrane and matrix material suggests that they may be involved in nutrient transfer.

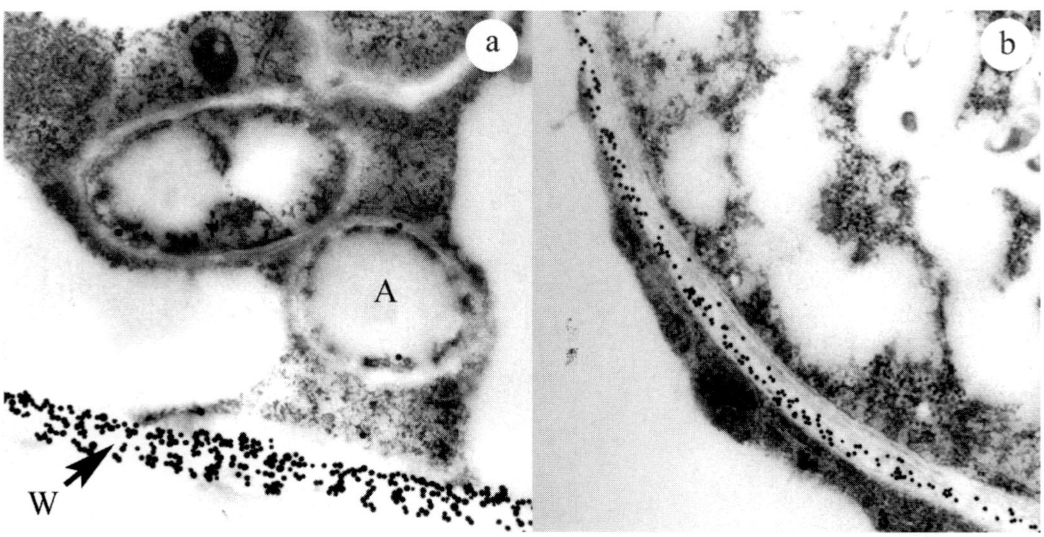

Figures a and b. Arbuscular mycorrhizas of *Zea mays* colonized by *Glomus versiforme* and labelled for cellulose using a cellobiohydrolase-gold complex. **Figure a.** Localization of cellulose in host cell wall (W). Only a few gold particles are localized around the small arbuscule branches (A). **Figure b.** Localization of cellulose in the interfacial matrix material around a large arbuscule branch. Both from Balestrini et al. Planta **195:** 201–209 (1994).

Box 10: Arbuscular mycorrhizal fungi harbour intracellular bacteria

As early as 1970, transmission electron microscopy showed bacterium-like organisms (BLOs) within spores and hyphae of arbuscular mycorrhizal (AM) fungi (Mosse 1970). Repeated observations of these in a number of AM fungal species was followed by attempts to isolate and culture the organisms as a means of identifying them and possibly determining their function as endosymbionts (Macdonald et al. 1982; Scannerini and Bonfante 1991). Although no one has succeeded in culturing them, molecular techniques enabled Bianciotto et al. (1996) to identify the bacteria within spores of *Gigaspora margarita* as pseudomonad bacteria similar to the genus *Burkholderia*. Amplified DNA from bacteria found in spores of *Scutellospora*, another member of the Gigasporaceae, using the same primers as for *G. margarita,* indicated that these bacteria may also be in the genus *Burkholderia*. In contrast, amplification of DNA from bacteria within spores of *Glomus mosseae* and *Acaulospora laevis* using these primers failed, suggesting that different bacteria may exist in these fungal species. Observations of the existence of intracellular bacteria similar to the genus *Burkholderia* have been extended to 11 fungal isolates within the family Gigasporaceae (Bianciotto et al. 2000). Only one isolate of *Gigaspora rosea* did not harbour bacteria. More recently Bianciotto et al. (2003) have published a new genus and species (*Candidatus Glomeribacter gigasporarum*) for the endosymbiotic bacteria that occur in arbuscular mycorrhizal fungi.

Using a specific fluorescent probe for bacteria, it has been shown that as many as 250,000 live bacteria occur in each *G. margarita* spore. Based on microscopic observations, it is known that BLOs appear to replicate within both spores and hyphae at various stages of root colonization (Bonfante et al. 1994).

Some progress has been made in determining possible functions of these bacteria. Nitrogen fixation genes have been identified in *Burkholderia* (Minerdi et al. 1998), and a putative P-transporter operon has been identified in the genome of *Burkholderia* living within the spore cytoplasm of *G. margarita* (Ruiz-Lozano and Bonfante 1999). The latter authors argue that the cost to the fungus and plant in terms of phosphate utilization by the bacteria may be outweighed by the gain in terms of nitrogen fixation.

With the diversity of prokaryotes, it is probable that more interactions with arbuscular mycorrhizal fungi will be discovered.

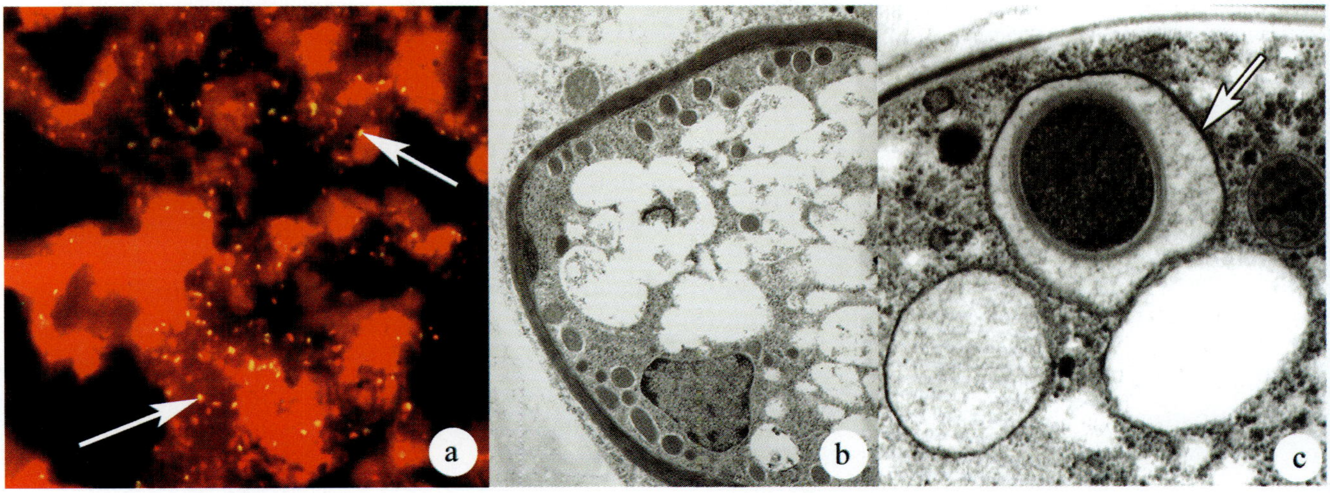

Figure a. Bacterial endosymbionts (arrows) in a spore of *Gigaspora margarita*. **Figure b**. TEM of bacterial endosymbionts in an intracellular hypha of *Glomus versiforme* in a leek root. **Figure c**. TEM of endosymbiotic bacteria within an intracellular hypha of *Gigaspora margarita* in a clover root. Each bacterium is surrounded by a membrane (arrow) of host origin. Figures a and c from Bianciotto et al. Appl. Environ. Microbiol. **62:** 3005–3010 (1996).

2. Functions

The highly branched nature of arbuscules and the presence of a periarbuscular membrane enveloping all of the fine branches provide a large surface area; this has led to the conclusion that it is through these structures that most of the nutrient exchange occurs. This conclusion will be strengthened by the demonstration that both sucrose and phosphate transporters occur in the periarbuscular membrane. It is known that each arbuscule formed in a root has a finite life span of a few days; therefore, it is possible that nutrients might also be released from the fungus into the root cell at the time of arbuscule degeneration.

D. Intraradical vesicles

1. Development and structure

As noted above, the genera *Gigaspora* and *Scutellospora* do not form intraradical vesicles. For those genera that do, there has been considerable interest in determining their role in the life cycle of the fungus. To address this, information related to their development and structure is important. Vesicles can develop from the tips of hyphae or from lateral branches, either within cells or in intercellular spaces of the root. When vesicles form within root cells, they frequently enlarge to occupy the entire volume of the cell; they may assume the shape of the cell (Figure 149) and even stretch the walls of the cells in which they occur (Figure 150). Little is known with respect to the early stages of vesicle development or the factors that trigger their formation. It is known that large numbers of vesicles may form in roots towards the end of the growing season, or in roots of pot cultures that have been maintained for long periods of time (Figure 151). Vesicle formation usually follows that of arbuscules, indicating that the fungus may require arbuscular-derived sugars from the host prior to their formation. Vesicle development involves the swelling of a hyphal tip or lateral branch followed by modifications to the cytoplasm and cell wall. The most characteristic cytological features of mature vesicles are the presence of lipid bodies (Figures 149, 150) and many nuclei. Bacteria also occur in vesicles of some species. The wall generally becomes thickened and this enhances the ability of vesicles to withstand soil drying and perhaps parasitism by soil organisms.

2. Functions

The storage of large amounts of lipid (as much as 58% of the dry mass of vesicles) indicates that vesicles are important storage structures and that they may act as propagules for the fungus. Declerck et al. (1998) isolated intact vesicles of *Glomus intraradices* (a species known to form a large number of intraradical vesicles) from genetically transformed roots of carrot, placed them on an agar medium, and allowed them to germinate. Germ tubes grew from the subtending hypha and these were able to contact and initiate colonization on adjacent roots. The use of vital stains has shown that vesicles can retain their viability even after three weeks of low temperature (5°C) storage, suggesting that these structures might be important over-wintering propagules. To determine, biochemically, the types of lipids present in vesicles, roots of colonized plants have been macerated and vesicles collected on a sucrose density gradient (Jabaji-Hare et al. 1984). Results indicated that of the three classes of lipids (neutral lipids, glycolipids and sphingolipids, phospholipids), glycolipids and sphingolipids were the most abundant followed by neutral lipids and phospholipids.

E. Extraradical vesicles (auxiliary vesicles)

1. Development and structure

Instead of intraradical vesicles, two genera, *Gigaspora* and *Scutellospora*, form auxiliary vesicles in the extraradical mycelium. These are initiated as lateral branches that rapidly expand into globose structures of varying colour and with ornamented walls (Figure 132). Auxiliary vesicles can occur singly or in clusters. Wall ornamentation patterns are used as taxonomic characters. Each auxiliary vesicle stores large amounts of lipid, has vacuolar inclusions, and is multi-nucleate. Bacteria may be present within the cytoplasm.

2. Functions

Very little is known about the role that auxiliary vesicles play in the life cycle of the fungus, although new hyphae have been observed growing from the subtending hyphae of these structures in *Gigaspora margarita*, suggesting that they may function in propagation.

F. Extraradical mycelium

1. Development and structure

The extraradical mycelium of arbuscular mycorrhizas, like that of other mycorrhizas, has been difficult to study *in situ* because of the heterogeneous nature of soil and the ability to locate the hyphae forming the mycelium network (Figure 152). Early studies involved extracting hyphae from the soil and examining these using light microscopy; extraradical hyphae comprising the mycelium network were described as dimorphic, having coarse hyphae with angular branchings as well as fine hyphae.

To resolve the problem of viewing extraradical mycelium in soil, glass-sided root chambers filled with soil, through which roots and fungal hyphae can grow, have been used (Friese and Allen 1991). Using this method, several types of mycelium were described. Hyphae that originated from germinating spores had a narrow diameter and, after contact with a root, repeatedly branched and developed several entry points. Other wide diameter hyphae originated from root fragments in the soil; these also often branched to form several entry points. Some hyphae (runner hyphae) had a wide diameter and usually did not branch as they contacted the root surface; these appeared to function mainly for the rapid spread of the fungus from adjacent living roots. Hyphae that exited the root were of two types. Wide diameter runner hyphae were identified that either grew along the surface of the root and established new entry points, or grew toward other roots and established hyphal bridges between adjacent roots of the same or different species. Other highly branched hyphae functioned presumably as the absorbing hyphal network.

Extramatrical mycelium development has also been studied in detail *in vitro* using genetically transformed roots inoculated with various fungal species (St. Arnaud et al. 1996; Bago et al. 1998*a,b*; Fortin et al. 2002) **(see Box 7)**. In these studies, hyphae originating from roots grew rapidly in a straight line across the agar surface and were mostly unbranched. Later, regions along the runner hyphae formed short, narrow diameter, lateral branches that in turn branched repeatedly to form 'tree-like' structures. Patches of branched hyphae, branched absorbing structures (BAS), formed at intervals as the runner hypha continued to grow across the agar surface (Bago et al. 1998*a*). Each patch lived 2–3 months under the study conditions. *In vitro* spore formation has occurred after the production of patches of absorbing hyphae and often in close proximity to these hyphae. Spore formation may rely on nutrients obtained via these hyphae. It is known from experiments by Giovannetti et al. (1994, 1996) that substances produced by host roots trigger branching of hyphae (Figure 153). In both soil-based and *in vitro* systems, these highly branched hyphae are likely the most important portion of the mycelium network in terms of nutrient uptake.

2. Functions

The extraradical mycelium has several functions, the most important of which is the uptake and translocation of mineral nutrients such as phosphorus from the soil solution to roots. The highly branched nature of the absorbing hyphae increases the surface area for nutrient uptake. Hyphal growth away from the depletion zone at the root surface extends the region from which nutrients and water can be absorbed (Figure 154). Increasing evidence suggests that the fine hyphae alter the pH in the adjacent soil resulting in increased availability of some nutrients. Hyphal bridges, known to occur between roots of adjacent plants, can act as a mechanism for the transfer of nutrients between hosts. The transfer of phosphate and carbon compounds is currently of considerable interest in terms of competition among plants.

An exciting new finding implicates hyphal bridges between some non-photosynthetic plant species and neighbouring photosynthetic species (Bidartondo et al. 2002). These authors used molecular methods to show that *Arachnitis uniflora*, five *Voyria* species, and *Voyriella parviflora* are associated with very specific AM fungi which were also present in neighbouring photosynthetic plants. It has yet to be shown, experimentally, that carbon compounds are transferred from the photosynthetic to the non-photosynthetic plants.

Figure 147. Hyphal coils of *Glomus intraradices* in cortical cells of a *Panax quinquefolius* root viewed with confocal laser scanning microscopy.

Figure 148. Arbusculate coils (arrowheads) of *Glomus intraradices* in cortical cells of a *Panax quinquefolius* root.

Figure 149. Vesicles of *Glomus versiforme* in cleared root of *Allium porrum*.

Figure 150. Sectioned vesicle of a *Glomus* sp. viewed with SEM. Large lipid bodies (*) are evident. Host cell wall indicated by arrowhead.

Figure 151. Large number of *Glomus versiforme* vesicles in an *Allium porrum* root.

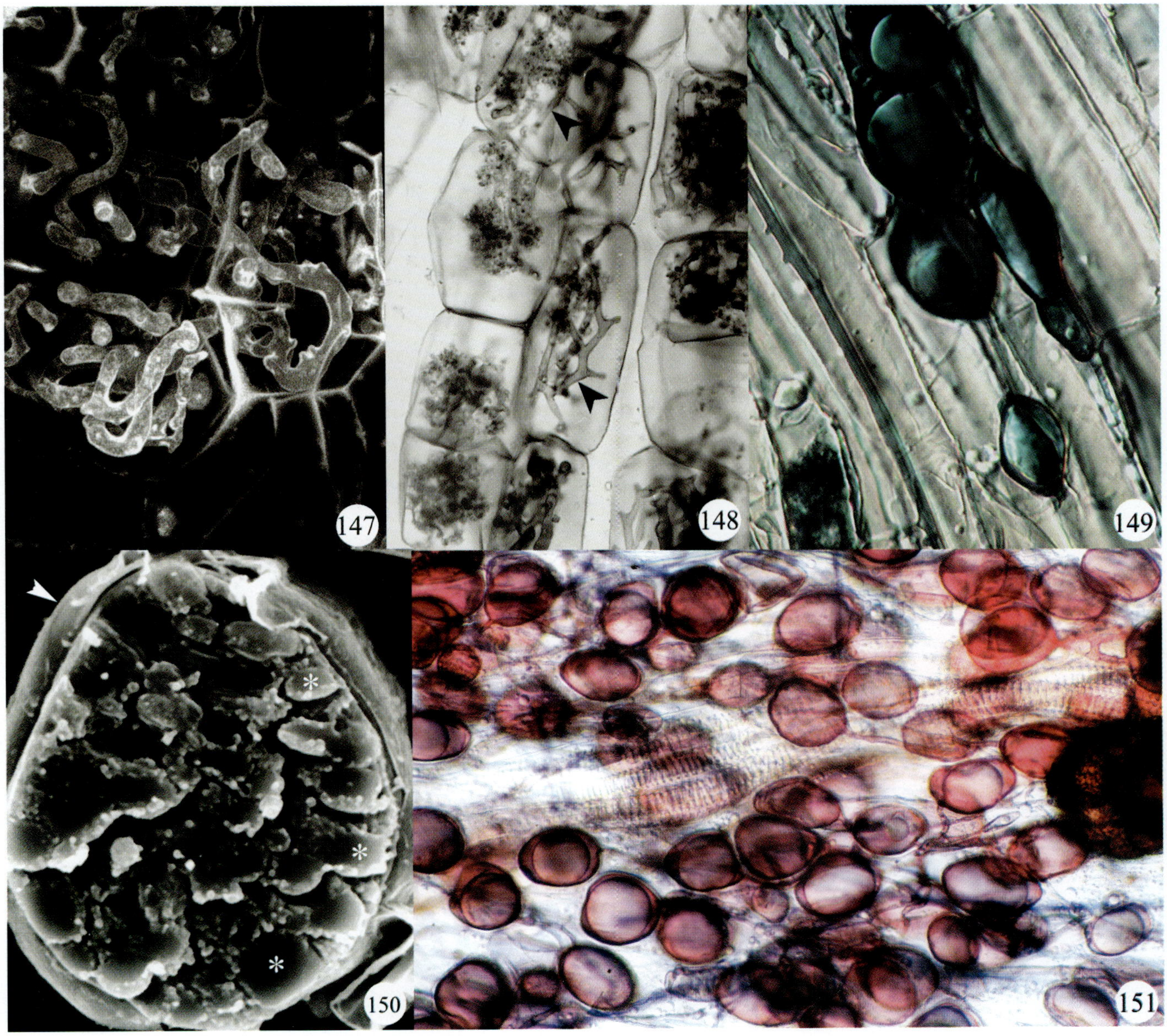
147
148
149
150
151

Figure. 152. Cleared *Allium porrum* roots colonized by *Glomus versiforme* showing extraradical mycelium (arrowheads) and developing spores (arrows).

Figure 153. Branching extraradical mycelium of *Glomus mosseae* separated from a root of *Alnus glutinosa* by a semi-permeable membrane. Photo from Giovannetti et al. New Phytol. **127:** 703–709 (1994).

Figure 154. Diagram showing the effect of extraradical mycelium of arbuscular mycorrhizas on the absorbing surface of the root. Hyphae (arrow) extend the region from which nutrients can be absorbed beyond the depletion zone of the root surface and root hairs (arrowhead).

Figure 155. Sand particles from a sand dune adhering to arbuscular mycorrhizal hyphae. Photo from Koske et al. Can. J. Bot. **53:** 87–93 (1975).

Figure 156. Hyphae of an arbuscular mycorrhizal fungus labelled with the fluorescent probe, FITC, conjugated with antibodies to glomalin, a glycoprotein produced by arbuscular mycorrhizal fungal hyphae. Fluorescence along the hyphae and within the soil indicates the presence of glomalin. Photo courtesy of Dr. Sara Wright.

152
153
154
155
156

The extraradical mycelium is also a source of inoculum for new roots. When the mycelium network is disturbed, either under laboratory or field conditions, there is a delay in colonization of new roots (Miller et al. 1995). This has important implications in agriculture in the way fields are prepared for seeding of new crops. Reduced tillage before seeding leads to the maintenance of a preexisting mycelium network and ensures that seedlings, once colonized, experience enhanced phosphorus absorption (Miller et al. 1995). Also, in various mining operations that involve considerable soil disturbance, reclamation of this soil often depends on planting pre-colonized plants since the mycelium network does not exist. The extraradical mycelium also may be destroyed by soil organisms that use it as a food source.

The extraradical mycelium aids in soil stabilization; soil particles adhere to the surface of hyphae (Figure 155) because of the production of a glycoprotein, glomalin (Figure 156). This process is important in the stabilization of sand dunes, soil in disturbed sites, and is likely important in maintaining soil texture generally. The extraradical mycelium plays an important role in the formation of spores that are important sources of inoculum for dispersal of the fungus (Figures 157–160).

G. Spores

1. Development and structure

Spore characteristics including size, colour, wall layers, and features of the subtending hypha are used to classify species of arbuscular mycorrhizal fungi. This has resulted in considerable information on the structure of mature spores, particularly features revealed by light microscopy (Figures 157, 158). Spores usually develop either singly as enlargements of the apex of extraradical hyphae or in groups within special hyphal aggregates (sporocarps) occurring in the extraradical mycelium (Figures 159, 160). In the genus *Paraglomus*, spores are initiated from the subtending hypha of an existing spore. Mature spores vary in the number of wall layers, the outer of which is often pigmented and impermeable. Wall layers are also of varying composition including polysaccharides, lipids, proteins and chitin. Chitin is present in one or more layers and often exists in a complex arrangement of fibrils. Characteristic cytological features include numerous storage lipid bodies, nuclei, and bacteria **(see Box 9)**. Spores do contain the usual organelles such as mitochondria, endoplasmic reticulum and vacuoles that are found in living fungal cells.

As propagules of arbuscular mycorrhizal fungi, spores need to germinate and produce hyphae that can colonize roots. The initial hyphae formed from resting spores are termed germ tubes. In many genera, germ tubes emerge from the subtending hyphal attachment; in others, they may emerge directly through the spore wall or from special regions that form within the spore. In those species in which germ tubes exit directly through the multi-layered wall it appears that enzymes digest the wall, thus enabling the germ tube to pass through.

2. Function

The formation of spores of arbuscular mycorrhizal fungi is affected by factors such as plant growth, light levels, and nutrients. Generally, more spores are formed in the extraradical mycelium towards the end of the growing season presumably after an extensive mycelium network has developed. A decline in spore production often occurs when photosynthesis of the host plant is reduced thereby limiting resource allocation to the root system and to the mycelium network. Spore production in some fungal species is influenced by the particular host species/fungal species combination.

The layered resistant wall of spores of most arbuscular mycorrhizal fungi enables them to persist in the soil, sometimes for many years. As new plant roots come into proximity with spores, they are induced to germinate and to form hyphae that colonize these roots. Although the spore wall is complex in structure, mycoparasitic fungi and bacteria frequently occur within spores and may render them non-viable.

Spore dispersal by air, water, and small animals is important in the distribution of arbuscular mycorrhizal fungi, especially with respect to severely degraded and disturbed soils, and agricultural soils that have been left either without crop cover or planted to non-mycorrhizal plant species for periods of time. These areas usually have very low spore numbers and the import of spores from adjacent areas becomes critical for the re-vegetation of degraded sites and for crop production on agricultural sites.

Box 11: Dynamic changes in the plant cytoskeleton

The cytoskeleton (microtubules and actin filaments) plays significant roles in the functioning of plant cells. For example, organelle movement, cell wall synthesis, and mitosis are major events involving the cytoskeleton. Among the changes that occur in root cells as arbuscular mycorrhizal associations develop, are dramatic alterations in the arrangement of the cytoskeleton (Timonen and Peterson 2002).With the entry of fungal hyphae into root cells and the subsequent development of arbuscules, changes in the organization of the cytoskeleton occur. In both *Arum*-type (Genre and Bonfante 1997, 1998; Matsubara et al. 1999; Blancaflor et al. 2001) and *Paris*-type (Armstrong and Peterson 2002) arbuscular mycorrhizas, the microtubules and actin filaments that are present in the peripheral cytoplasm of root cells prior to fungal invasion largely disappear to be replaced by new populations of both of these elements. Microtubules and actin filaments become closely associated with hyphae and arbuscules.

Blancaflor et al. (2001) have shown that root cells adjacent to those that have been colonized by intracellular and intercellular hyphae also show a rearrangement of microtubules. This suggests that host cells receive a signal from either adjacent root cells or from the fungus prior to invasion, triggering changes in the cytoskeleton. As arbuscules degenerate, the cortical microtubules and actin filaments reappear in root cells.

The functions of microtubules and actin filaments during colonization are not clear but they may be involved in the positioning of the periarbuscular membrane, in nuclear movement, and the movement of precursors for the formation of the interfacial matrix between the periarbuscular membrane and the fungal cell wall. The demonstration that microtubule organizing centres and the protein, clathrin, are both present in the periarbuscular membrane suggests that clathrin-coated vesicles might be involved in the formation of the interfacial matrix (Genre and Bonfante 1999).

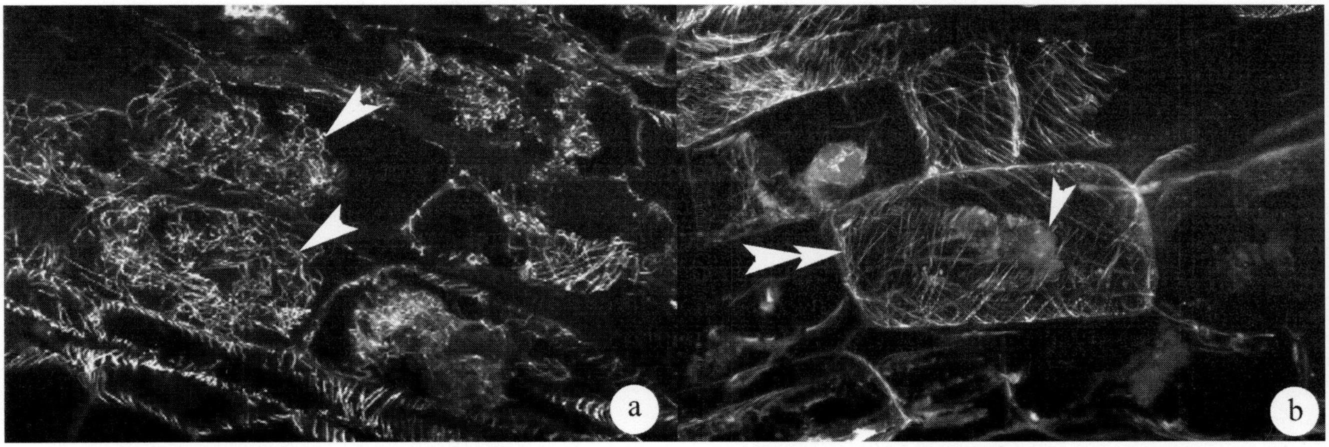

Figures a and b. Cortical cells of *Asparagus officinalis* colonized by *Glomus intraradices* using immunocytochemistry to localize microtubules. **Figure a.** Microtubules are closely associated with arbuscule branches (arrowheads). **Figure b.** A collapsed arbuscule (arrowhead) is associated with a few microtubules and cortical microtubules (double arrowhead) have reappeared.

Figures 157, 158. Spores of *Gigaspora calospora* showing large lipid bodies (arrowheads) and the subtending hypha (arrow).

Figure 159. Sporocarp of a *Glomus* sp. from a maple – beech forest. Spores (*) and hyphae (arrowheads) are evident. Photo courtesy of Dr. Mark Brundrett.

Figure 160. Sporocarp of *Glomus intraradices* containing many spores (arrows) from a pot culture of *Allium porrum*.

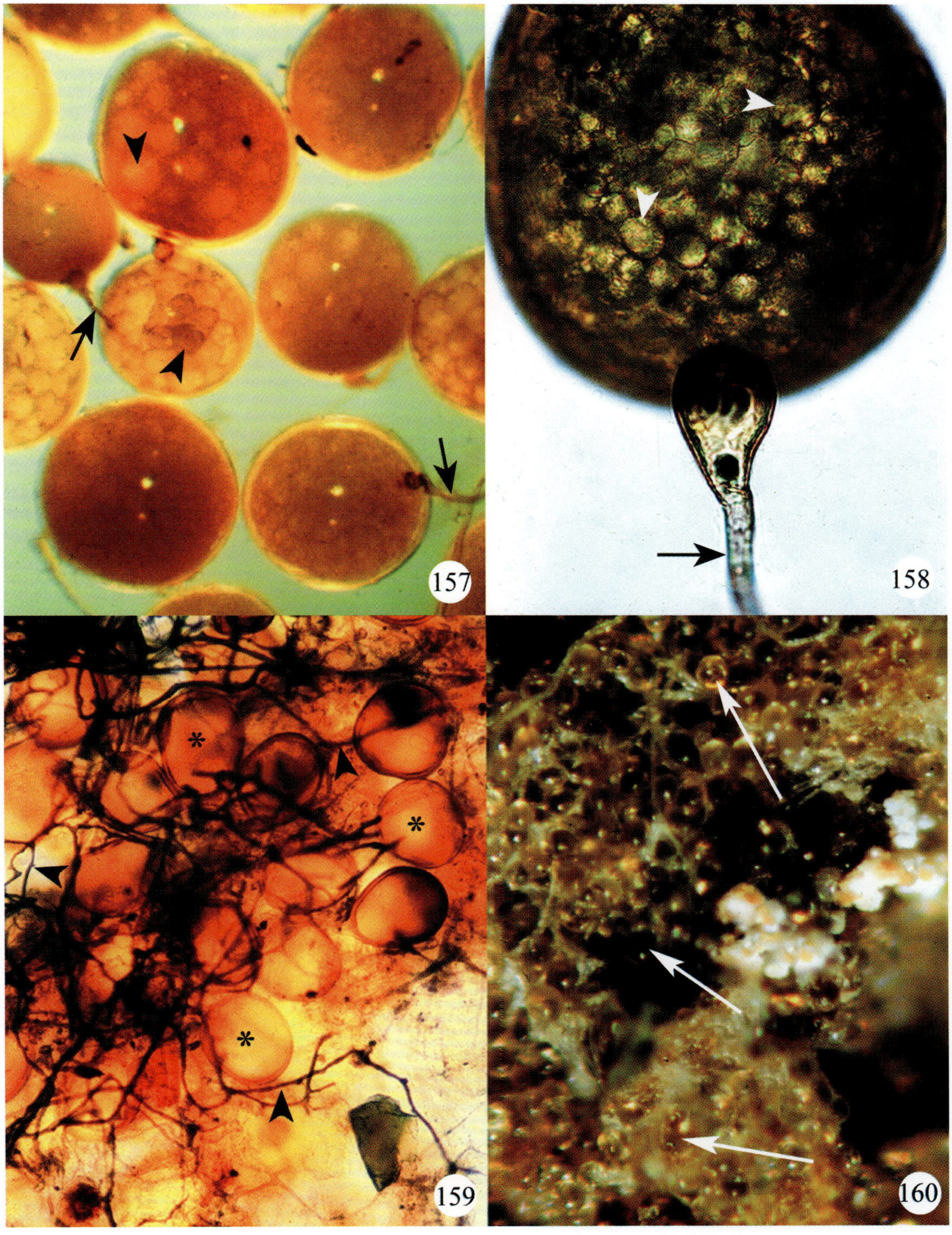
157
158
159
160

Chapter 4. Ericoid mycorrhizas

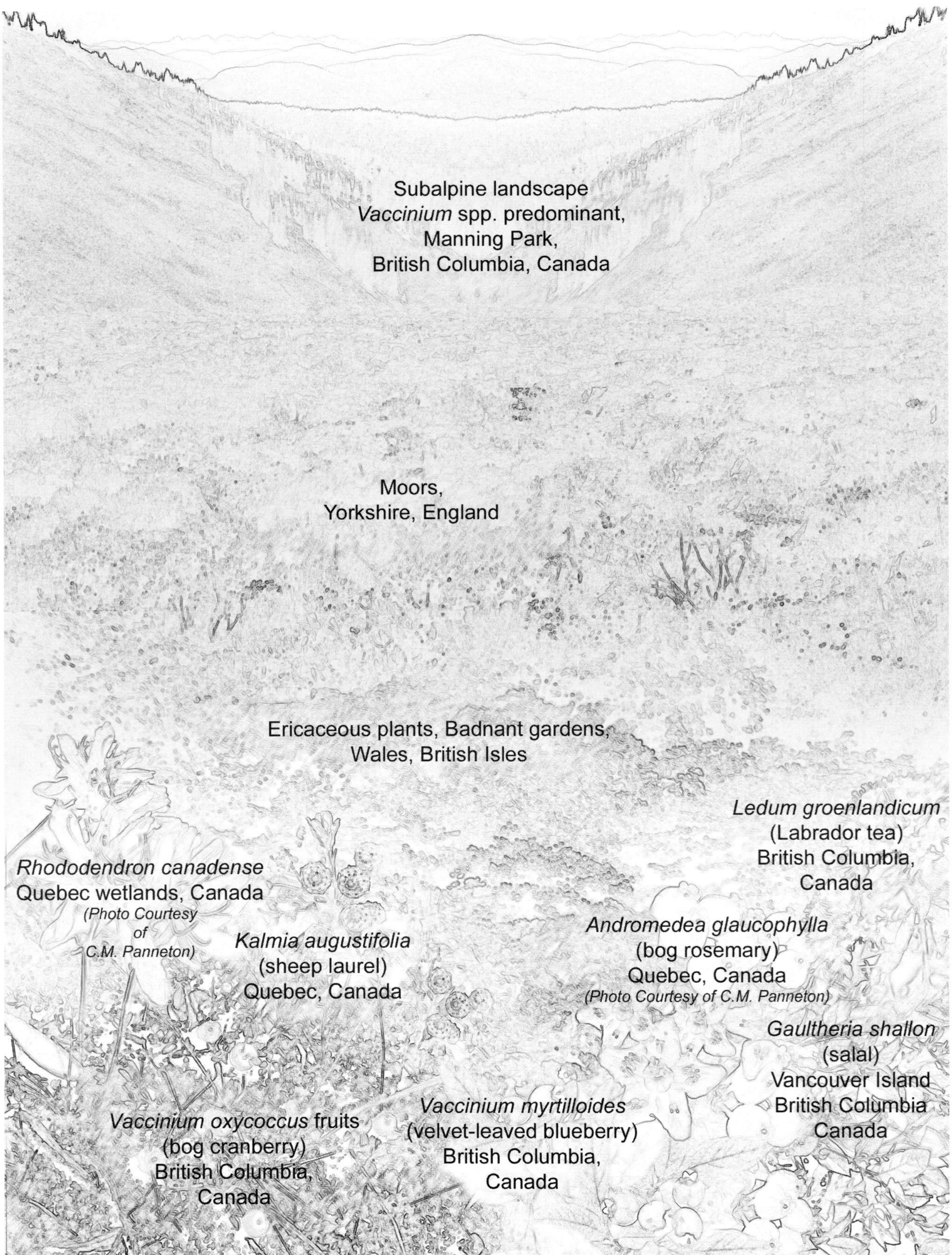

Subalpine landscape
Vaccinium spp. predominant,
Manning Park,
British Columbia, Canada
Moors,
Yorkshire, England
Ericaceous plants, Badnant gardens,
Wales, British Isles
Ledum groenlandicum
(Labrador tea)
British Columbia,
Canada
Rhododendron canadense
Quebec wetlands, Canada
(Photo Courtesy of C.M. Panneton)
Kalmia augustifolia
(sheep laurel)
Quebec, Canada
Andromedea glaucophylla
(bog rosemary)
Quebec, Canada
(Photo Courtesy of C.M. Panneton)
Gaultheria shallon
(salal)
Vancouver Island
British Columbia
Canada
Vaccinium myrtilloides
(velvet-leaved blueberry)
British Columbia,
Canada
Vaccinium oxycoccus fruits
(bog cranberry)
British Columbia,
Canada

Chapter 4. Ericoid mycorrhizas

A. Introduction

1. Definition

Ericoid mycorrhizas represent a unique type of mycorrhizas confined to several families in the large angiosperm order Ericales. A unifying feature of plants developing this type of mycorrhiza is the formation of very specialized lateral roots, 'hair roots' (Figure 161). These are very narrow in diameter, possess a simple anatomy, have limited extension growth, and lack secondary growth. Each root consists of a narrow vascular cylinder, one or two rows of cortical cells (including the endodermis), and an epidermal layer of enlarged cells (Figure 162). The mycorrhizal association involves the colonization of epidermal cells by fungal hyphae followed by the formation of a branched hyphal complex in each colonized cell (Figure 163).

Although ericoid mycorrhizas are confined to the order Ericales, species in two main families, Ericaceae and Epacridaceae, contribute significantly to ecosystems in the Northern and Southern hemisphere respectively. In the Northern hemisphere, Ericaceae species with typical ericoid mycorrhizas often dominate heathlands; they are also well represented in the subalpine and alpine floras of both Europe and North America. Acidic heathlands and sandplains in the Southern hemisphere (notably in Australia) are often dominated by species in the family Epacridaceae (Cairney and Ashford 2002).

In all situations plants with ericoid mycorrhizas are found growing on nutrient-poor soils, suggesting that ericoid mycorrhizal associations confer an important function by increasing the capacity of these hosts to absorb mineral nutrients.

In western North America, the ericaceous species *Gaultheria shallon* (salal) often proliferates in areas after forests have been cut or burned; the dense growth of salal sometimes results in delaying conifer seedling establishment. In Europe, an array of ericaceous species have colonized vast tracts of heathland.

Few genera in the Ericaceae are considered to be commercially important; however, those that are, contribute significantly to the economies of the areas in which they grow. Shoots of *G. shallon*, harvested for floral arrangements in British Columbia, Canada, represent a 45–60 million dollar business every year. Various species of *Vaccinium* (the blueberries and cranberries) are grown for their fruit while the shoots of some *Vaccinium* species are used in the florist trade. Other ericaceous species, including several in the family Epacridaceae, are grown as ornamentals.

2. Plant species involved

The order Ericales consists of 25 families with approximately 9,450 species; however, species in only three families (Ericaceae, Epacridaceae and Empetraceae) usually possess typical ericoid mycorrhizas. Although molecular studies have placed the Epacridaceae and Empetraceae in the family Ericaceae (Judd et al. 1999), we will discuss them as separate families in this book. In Hawaii, three species in the genus *Vaccinium* (Ericaceae) as well as the species *Styphelia tameiameiae* (Epacridaceae) have arbuscular mycorrhizas instead of ericoid mycorrhizas (Koske et al. 1990). The genera *Arbutus* and *Arctostaphylos*, both in the Ericaceae, and most members of the family Pyrolaceae have specialized mycorrhizas known as arbutoid (see Chapter 5); plants in the family Monotropaceae have monotropoid mycorrhizas (Chapter 6).

Many ericaceous families including Clethraceae, Grubbiaceae and Cyrillaceae have been poorly studied. Roots of *Clethra barbinervis* (Clethraceae) collected from the field and roots of seedlings grown in the greenhouse in soil collected near a tree of this species developed the *Paris*-type arbuscular mycorrhiza (Kubota et al. 2001). Overall, the mycorrhizal status of many species in the order Ericales remains undescribed.

The immense diversity within the order Ericales and its worldwide distribution makes characterizing plants having ericoid mycorrhizas very difficult. In general terms, these plants are often perennial shrubs or small trees with sclerophyllous (i.e., highly lignified) leaves. There is a wide range of flower types in this order and many species in the Epacridaceae have highly modified flowers for bird pollination.

Figure 161. Diagrammatic representation of an ericoid species with numerous hair roots (arrowheads).

Figure 162. Diagram of a transverse section of a hair root showing enlarged epidermal cells (E) with a thickened outer tangential wall, one layer of cortex (C), an endodermis (En), a few xylem tracheary elements (X), and a few phloem cells (P).

Figure 163. Diagram of epidermal cells showing entry of fungal hyphae through the thickened wall (arrowhead), hyphal complexes (HC), and a narrow hypha (double arrowhead) connecting adjacent epidermal cells.

161

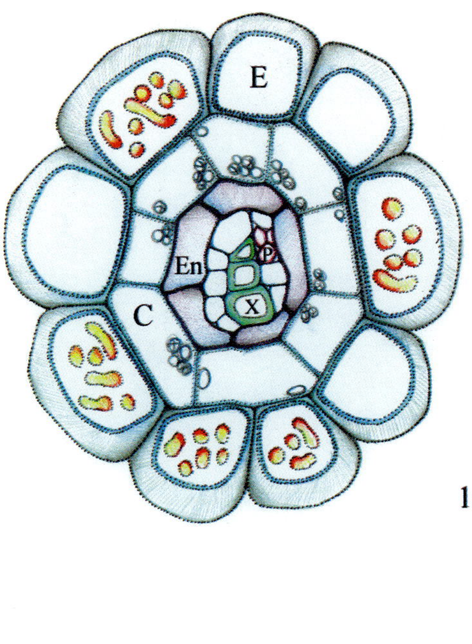

162

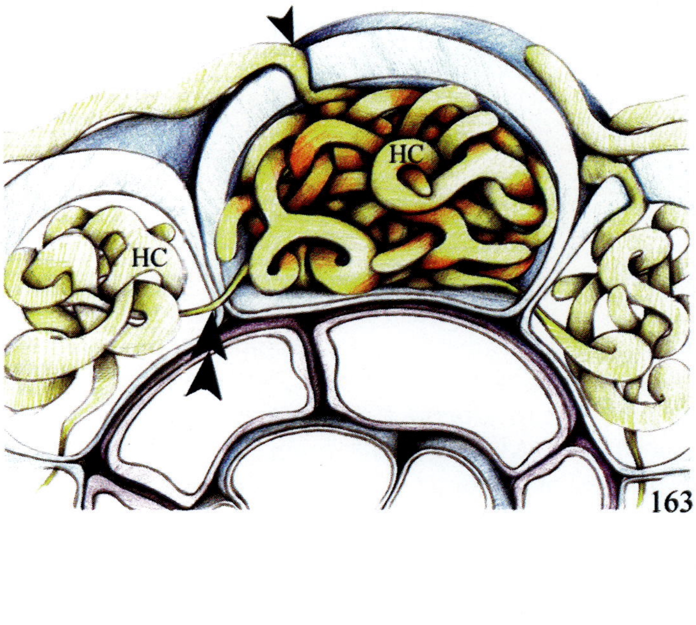

163

In an interesting study, Duckett and Read (1995) have shown that isolates of *Hymenoscyphus ericae*, the most common ericoid fungus, can colonize rhizoids of genera in several leafy liverwort families and, conversely, that cultures of fungi isolated from several leafy liverworts can form typical ericoid mycorrhizas with several ericaceous species. In the liverworts, *H. ericae* triggers the enlargement of the rhizoid tip within which a hyphal complex develops **(See Box 12)**.

3. Fungal species involved

A diverse assemblage of fungi has been isolated from hair roots of a number of genera within the Ericaceae and Epacridaceae. Isolates are generally slow-growing and many produce dark mycelium (dematiaceous mycelium) when grown on various media (Figure 164). For many years, only one fungal species, *Hymenoscyphus ericae* = *Pezizella ericae* (Leotiales), identified originally by characteristics of the mycelium in culture and the production of asexual propagules (arthroconidia, Figure 165), had been shown by synthesis experiments to form ericoid mycorrhizas with members of the Ericaceae (Smith and Read 1997). The characteristic arthroconidia place this fungus in the anamorph genus, *Scytalidium*. Cultures of the fungus have formed sexual reproductive structures (apothecia with asci and ascospores), and molecular evidence confirms that at least one isolate of *Scytalidium (S. vaccinii)* is the anamorph of the teleomorph, *H. ericae*.

Several other isolates are known to form typical ericoid mycorrhizas with a variety of ericaceous hosts (see review by Berch et al. 2002). Among these, anamorph species of *Oidiodendron* and their teleomorphic states, *Myxotrichum* and *Byssoascus* (Hyphomycetes), are especially important. Many unidentified isolates cultured from ericaceous hosts also form typical ericoid mycorrhizas; their taxonomic placement is beginning to be clarified with the help of molecular methods **(see Box 13)**.

A dark septate fungal endophyte, *Phialocephala fortinii*, was isolated from the roots of all ericaceous species examined from an alpine heath site and a sand dune site, and from two of five species from a bog site in Alberta, Canada (Hambleton and Currah 1997). Although this fungal species and other dark septate fungal endophytes do not form typical ericoid mycorrhizas, their prevalence in roots of many species in the Ericaceae (Figures 166, 167) is worth noting and deserves further study (see Chapter 8).

Many fungal isolates have also been obtained from hair roots in the Epacridaceae. The mycelium of many, in culture, is dematiaceous, but isolates can range in colour from white to pink to various dark shades. Molecular studies show genotype similarities between some of the Epacridaceae isolates, *Hymenoscyphus, Oidiodendron*, and other isolates from the Ericaceae (Berch et al. 2002). As more host species in both families, Ericaceae and Epacridaceae, are examined with respect to their fungal partners, there is no doubt that the number of identified fungal species will increase. There is some evidence that fungal species belonging to the basidiomycetes may also be involved in forming ericoid mycorrhizas.

B. Development and structure

The surface of hair root tips is surrounded by mucilage (Figures 168, 169) that potentially harbours microorganisms and in which hyphae of ericoid mycorrhizal fungi may proliferate (Figure 169) before colonizing epidermal cells. The mucilage is secreted by the root cap and differentiating epidermal cells and is composed of sugars, N-acetylglucosamine, polygalacturonic acid and β–1,4 glucans (Peretto et al. 1990). It is not known whether this mucilage participates in recognition between fungus and root or whether it simply provides a niche for the growth of fungi and other organisms. Fungal hyphae also secrete a fibrillar sheath containing glucose and mannose sugar residues (Perotto et al. 1995). This sheath is more pronounced in infective fungal strains compared to non-infective strains, suggesting that components of the sheath are involved in recognition between symbionts. The sheath may also aid in the attachment of hyphae to the surface of the root; once a hypha enters the epidermis, the sheath disappears.

Hyphae of fungal species that can form ericoid mycorrhizas must contact the surface of hair roots and then penetrate the epidermal cell wall (Figures 170, 174) before colonizing epidermal cells and establishing hyphal complexes (Figures 170, 171). Although it has been widely accepted that each

Box 12: Liverworts and ericoid species can share common fungal symbionts

An excellent discussion on the history of reports of liverworts being colonized by fungi can be found in the review by Read et al. (2000). A remarkable finding is that some members of leafy liverworts can associate with members of the Glomeromycota and form arbuscules and vesicles, structures formed in arbuscular mycorrhizal associations in vascular plants. Other species associate with basidiomycetes, and still others with ascomycetes. In the latter case, using a fluorescent probe specific to this group of fungi (Duckett and Read 1991), and ultrastructural studies in which simple septa with Woronin bodies were demonstrated in hyphae (Duckett et al. 1991), ascomycete fungi have been localized in the shaft and swollen tip of rhizoids in several leafy liverwort families. Duckett and Read (1995) later demonstrated, experimentally, that the ascomycete fungus *Hymenoscyphus ericae,* a species that forms typical ericoid mycorrhizas with a range of ericoid plant species, colonized rhizoids of several families of leafy liverworts and induced tip swelling. Swollen rhizoid tips are reported in these liverwort families in nature. These authors also showed that fungi isolated from colonized rhizoids of field-collected liverwort species, formed typical ericoid mycorrhizas with seedlings of *Erica cinerea*, *Calluna vulgaris*, and *Vaccinium corymbosum*. The physiological role of liverwort-fungal associations has yet to be determined.

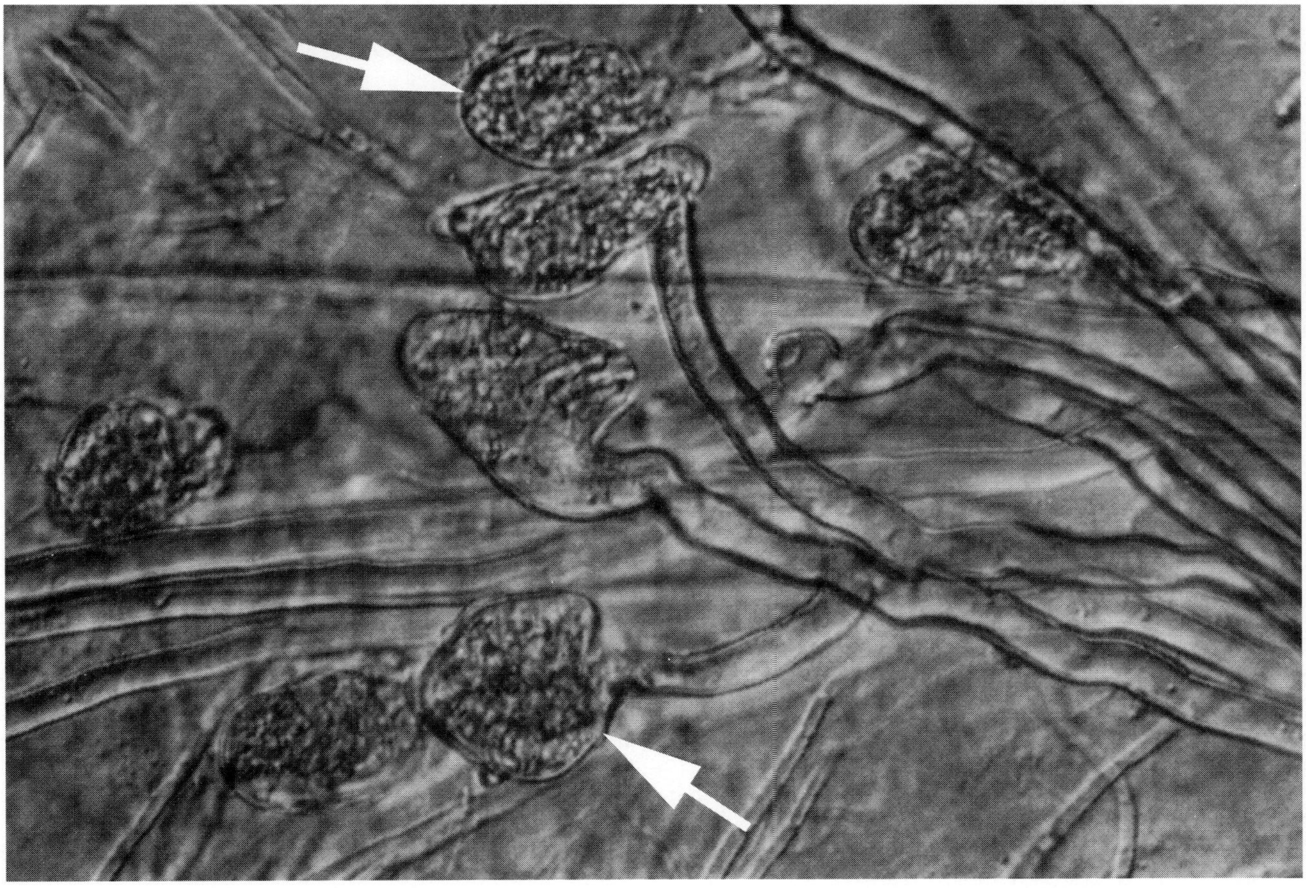

Swollen rhizoid tips of the liverwort, *Calypogeia muellerana,* colonized by the ericoid fungus *Hymenoscyphus ericae*. From Duckett and Read. New Phytol. **129:** 439–447 (1995).

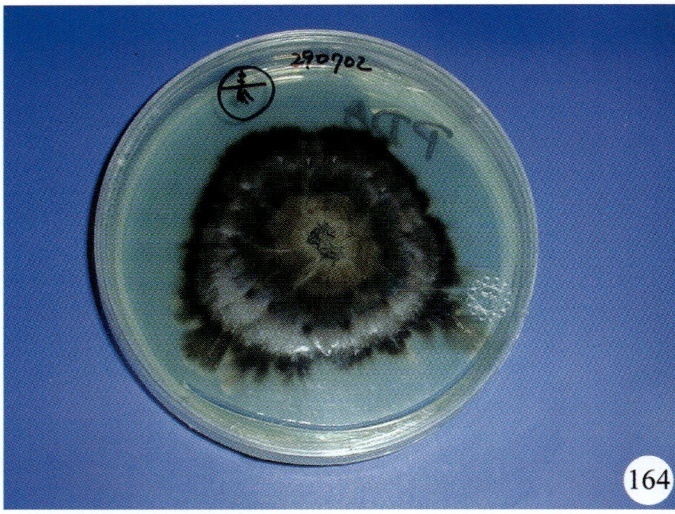

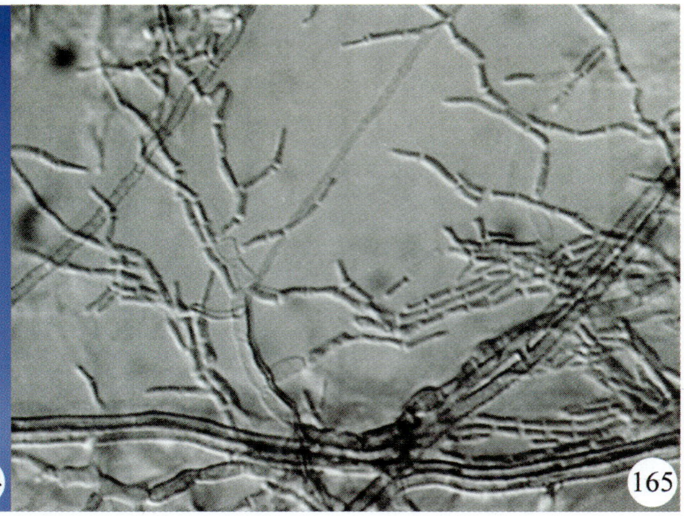

Figure 164. Culture of *Hymenoscyphus ericae*.

Figure 165. Arthroconidia (asexual propagules) of *Scytalidium vaccinii* (the anamorph state of *Hymenoscyphus ericae*). Photo from Hambleton and Currah. Can. J. Bot. 75: 1570–1581 (1997).

epidermal cell is invaded by a separate infection hypha without lateral spread from epidermal cell to epidermal cell, recent evidence in our lab indicates that this may not always be true (Figures 172, 173). Many species in the Ericaceae and Epacridaceae have very thick epidermal cell walls (Figure 170) through which fungal hyphae must pass in order to enter the root. In the case of *Lysinema ciliatum* (Epacridaceae), the thickened wall is multi-layered and may function either as a source of nutrients for the enclosed fungus or perhaps to protect the hyphal complex during periods of drought (Ashford et al. 1996). These authors also suggest that sloughed thick-walled epidermal cells containing hyphal complexes may act as propagules for the fungus.

Observations from field-collected material and from seedlings grown *in vitro* indicate that the symbiosis between ericoid mycorrhizal fungi and hair roots, regardless of the fungal and plant species involved, share some common characteristics. The colonization of roots is always restricted to epidermal cells and involves the coiling and branching of a fungal hypha to form a hyphal complex within each cell (Figures 170, 174). Not all epidermal cells of a root become colonized, but in many field collections, colonization levels can be very high (Figure 175). Fungal hyphae within epidermal cells are surrounded by host plasma membrane (Figure 176). An interfacial matrix is present between the fungal cell wall and host plasma membrane but it is simpler in composition compared to that described for arbuscular mycorrhizas. Labelling with antibodies specific for host cell wall components has shown that only arabinosyl residues are present (Perotto et al. 1995). Nevertheless, movement of nutrients between the symbionts must occur across this membrane-interfacial matrix boundary.

Epidermal cells containing hyphal complexes have an enlarged nucleus and are rich in organelles including mitochondria, plastids, and components of the endomembrane system. Fungal hyphae typically have dolipore septa with pronounced associated Woronin bodies (indicative of their ascomycete affiliation), and frequently contain deposits of glycogen and polyphosphate. Organelles such as nuclei, mitochondria and ribosomes are present in hyphae. The longevity of any epidermal cell-hyphal complex is variable. Degradation of the host and fungal cytoplasm occurs in both field-collected material and in *in vitro* systems. The breakdown of the symbiotic association may partly depend on the longevity of an individual hair root. It is generally accepted that there is rapid turnover of these fine roots in soil.

The extent of development of surface hyphae on hair roots is variable (Figures 171, 177); in some cases a mantle-like structure forms (Figure 178). Many of the hyphae on hair root surfaces are runner hyphae from which lateral branches develop that enter epidermal cells (Read 1984).

C. Extraradical mycelium

The extent of the extraradical mycelium in ericoid mycorrhizas is very limited and this may be due to the fact that the extensive hair root system plays the major role in soil exploitation (Read 1984). This author reports that it is difficult to extract mycelium from peat soil and therefore there are currently no good measurements of the extent of the extraradical mycelium from the roots of ericoid plants in such soils. However, measurements have been obtained in sandy heathland soil; here, extraradical mycelium does not extend beyond 1 cm from the hair root surface and most extraradical hyphae are located on and immediately adjacent to the root surface (Read 1984).

D. Functions

The prevalence of ericoid mycorrhizas in nutrient-poor heathlands, in both the northern and southern hemispheres, has stimulated research related to their function. Read and colleagues (see Read 1996; Smith and Read 1997) have intensively studied the roles of ericoid mycorrhizas in northern hemisphere heathlands. Fewer studies have examined the function of epacrid mycorrhizas in the southern hemisphere. Depending on soil conditions, ericoid mycorrhizas can have several roles. It has been demonstrated that they can take up phosphate, an important role of most other mycorrhizas. Hyphae of *Hymenoscyphus ericae*, the most prevalent ericoid mycorrhizal fungus, can secrete acid phosphatases onto the surface of the fibrillar sheath surrounding external hyphae that grow close to host roots. These enzymes enable the host plant to access phosphate from organic and (or) condensed phosphates.

The main role of ericoid mycorrhizas, at least in heathlands, may involve the acquisition of nitrogen which can be limiting to plant growth in these areas **(see Box 14)**. Ericoid fungi are able to obtain nitrogen from a number of sources. Nitrate and ammonium ions, and free amino acids can be absorbed by hyphae and translocated to the host. In addition, nitrogen can be obtained from complex organic substrates through the production of proteinases, and from chitin, the major structural component of most fungal cell walls, through the production of chitinases. These sources of nitrogen would be unavailable to plants without their mycorrhizal associations. Much of the nitrogen in soils supporting ericoid plant species is bound in organic matter; ericoid mycorrhizal associations play a major role in the success of ericoid species in accessing this nitrogen. In addition, ericoid fungi can degrade pectins and lignins, components of plant cell walls, making carbon compounds available for fungal growth (and perhaps plant growth) during times of limited photosynthetic activity. The hair roots of ericoid plants occur predominantly in the superficial layers of soil rich in plant litter and microorganisms. This places them, and their associated mycorrhizal fungi, in a favourable position to scavenge nitrogen, carbon and phosphorus from organic materials.

Ericoid mycorrhizal fungi may also have a significant role in terms of mobilization and uptake of iron when it is present at low concentration or low availability. The production of a hydroxamate siderophore may be involved in the ability of these fungi to access iron under such conditions.

In certain environments, ericoid mycorrhizal fungi have the ability to protect their hosts from toxic levels of heavy metals such as copper and zinc, two metals often associated with mine spoils and found at high levels in soils adjacent to smelters (Bradley et al. 1982). Strains of ericoid mycorrhizal fungi have been isolated from soils contaminated with cadmium and arsenic; these fungi tolerate higher levels of these metals when grown *in vitro* compared to strains from non-polluted sites (Perotto et al. 2002). The slime (mucilage) produced by hyphae of *Hymenoscyphus ericae* is important in taking up zinc, binding it, and thus decreasing zinc content in the shoots of host plants (Denny and Ridge 1995). As well, there is the possibilty that hyphae within root epidermal cells may concentrate heavy metals, preventing them from being transported to the shoot where they could interfere with metabolic processes.

The results to date show that ericoid mycorrhizas play a significant role in the success of many members of the Ericales in soils that are nutrient deficient or contaminated with metals. One concern expressed by Read (1996) is that ericaceous species, adapted to soils with low levels of available nitrogen, may lose this advantage in regions of high deposition of nitrogen from pollution. He suggests that evidence for this already exists in the heathlands of northwest Europe where grasses are replacing ericaceous species.

Figure 166. Dark septate fungal endophytes (arrowheads) in cleared roots of *Gaultheria shallon*.

Figure 167. Microsclerotia (arrowheads) of dark septate fungal endophytes in cleared roots of *Vaccinium* sp.

Figure 168. Hair root of *Gaultheria shallon* stained with toluidine blue O showing the mucilage sheath (*) and trapped sloughed root cap cells (arrowheads).

Figure 169. PAS-stained root tip of *Gaultheria shallon* counterstained with alcian blue showing root cap (rc), mucigel (*) and fungal hyphae of *Hymenoscyphus ericae* (arrows).

Figure 170. Cleared root of *Gaultheria shallon* showing thickened epidermal cell walls (arrowheads) and intracellular fungal coils (*).

Figure 171. Transverse section of *Gaultheria shallon* hair root showing a mantle-like structure (m), a hypha penetrating the epidermis (arrow), and hyphal coils (arrowheads).

Cultured seedlings of *Gaultheria shallon* inoculated with *Hymenoscyphus ericae* used for Figures 168, 169 were provided by Dr. Shannon Berch.

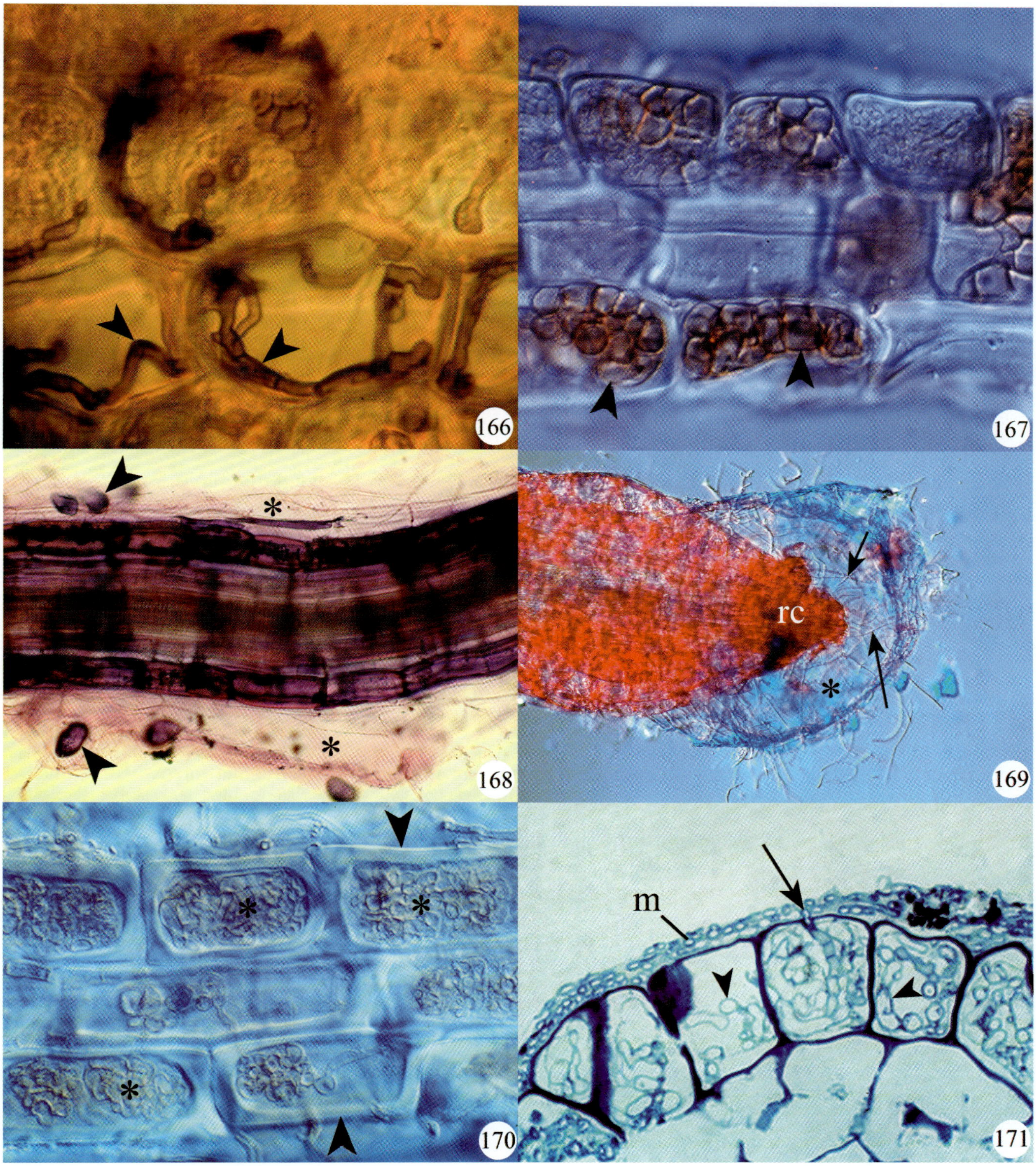
166
167
168
rc
169
170
m
171

Figure 172. Cleared root of *Vaccinium myrtilloides* stained with acid fuchsin and viewed with confocal laser scanning microscopy. Narrow hyphae (arrowheads) connect hyphal complexes in adjacent epidermal cells.

Figure 173. Cleared root of *Ledum groenlandicum* stained with acid fuchsin and viewed with confocal laser scanning microscopy. Hyphal complexes in adjacent cells are connected by narrow hyphae (arrowheads).

Figure 174. Well-developed hyphal complexes in epidermal cells of a *Vaccinium oxycoccus* hair root. Root was cleared, stained with acid fuchsin, and examined with confocal laser scanning microscopy.

Figure 175. Field-collected hair root of *Gaultheria procumbens* showing high colonization level. Root cleared, stained with acid fuchsin, and examined by confocal laser scanning microscopy.

Figure 176. Transmission electron micrograph showing a sectioned hyphal coil (*) within a hair root epidermal cell of *Calluna vulgaris*. The coil is separated from the host cell cytoplasm by a plasma membrane (arrowheads). Photo from Perotto et al. Can. J. Bot. **73 (Suppl. 1):** S557–568 (1995).

Figure 177. Section of *Ledum groenlandicum* showing limited surface hyphae (arrowheads) and hyphal complexes (arrows) in some epidermal cells.

Figure 178. Surface view of *Gaultheria procumbens* root showing the development of a mantle-like hyphal covering.

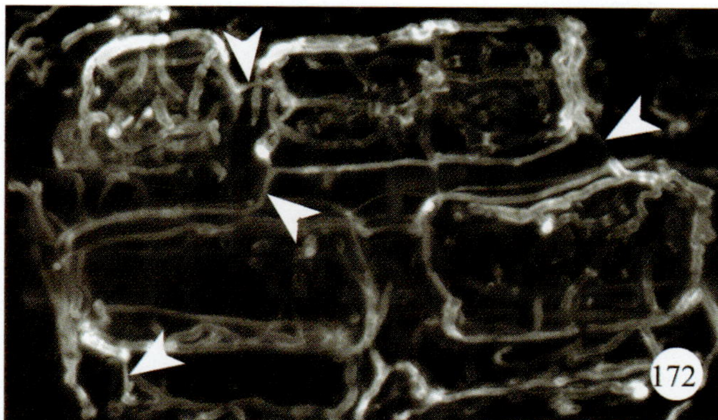
172

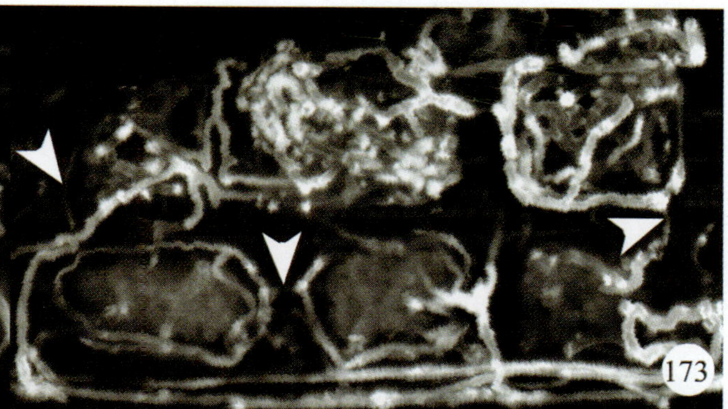
173

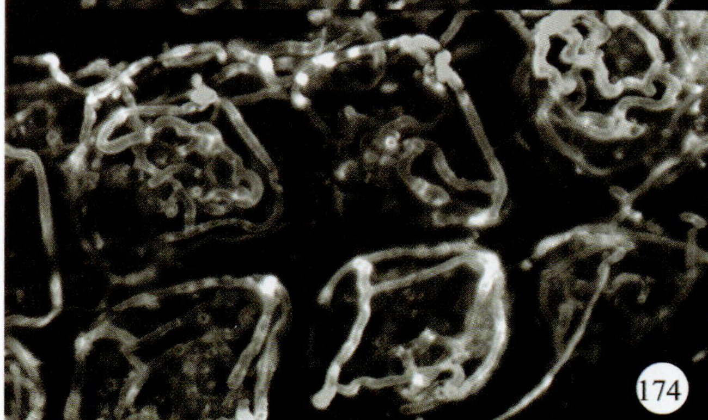
174

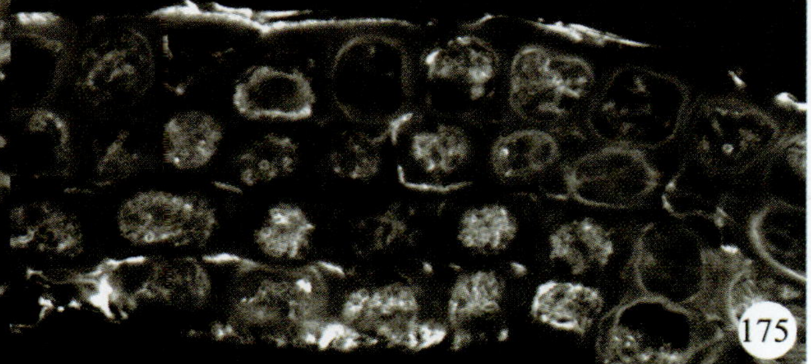
175

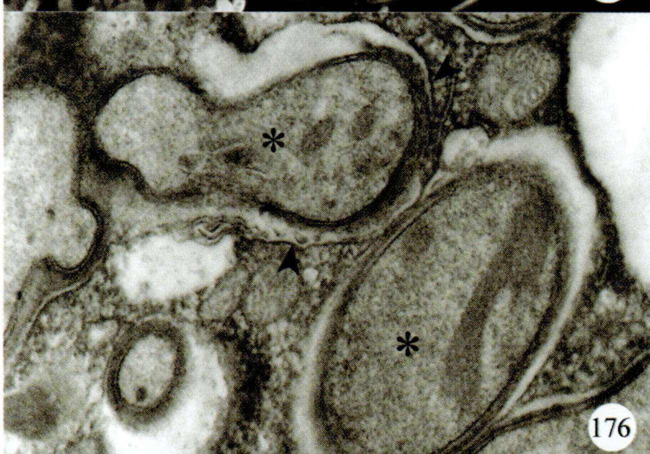
176

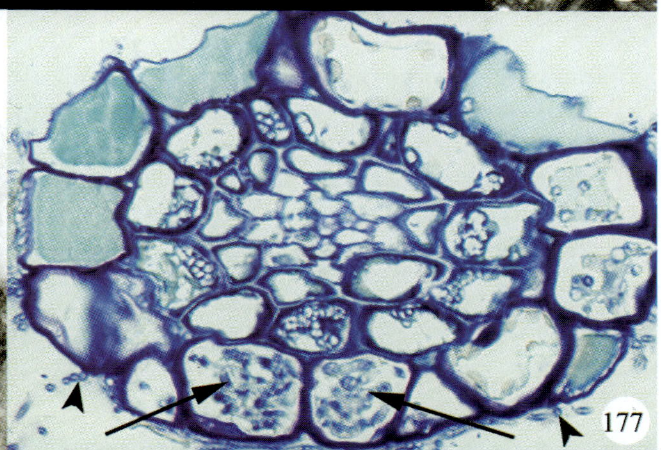
177

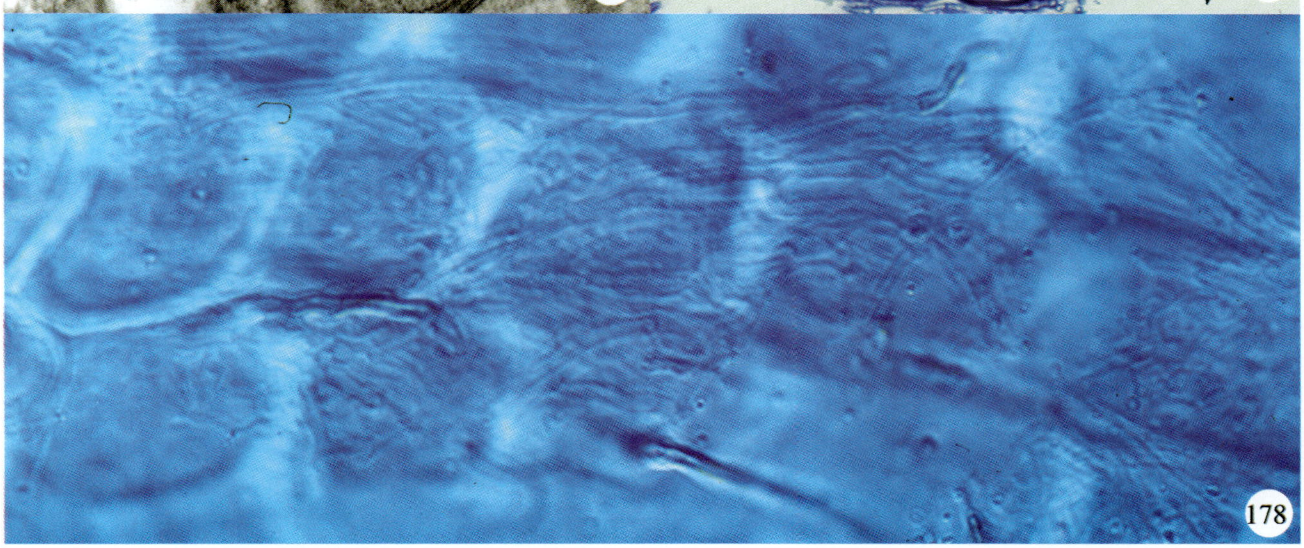
178

Box 13: Investigating the identity of ericoid mycorrhizal fungi

Unlike the other major mycorrhizal categories, few fungal species (and these are mostly ascomycetes) have been shown, experimentally, to form typical ericoid symbiotic associations. Until 1983, only two fungi, *Hymenoscyphus ericae* = *Pezizella ericae* (identified by Read 1974, as *P. ericae* and later renamed *H. ericae* by Kernan and Finocchio 1983), and *Oidiodendron griseum* (Burgeff 1961) were reported to form ericoid mycorrhizas. Couture et al. (1983) confirmed that *O. griseum* formed typical ericoid mycorrhizas with *Vaccinium angustifolium* and since then, several other *Oidiodendron* species have been shown to form this type of mycorrhiza (Xiao and Berch 1995; Hambleton et al. 1998; Monreal et al. 1999). Other species with *Oidiodendron* anamorphs (e.g., *Myxotrichum setosum* and *Gymnascella dankaliensis* (Dalpé 1989)) and *Pseudogymnoascus roseus* with a *Geomyces* anamorph (Dalpé 1989) have been confirmed to form typical ericoid mycorrhizas. The anamorph species, *Acremonium strictum* (Xiao and Berch 1996) and *Stephanosporium cerealis* (Dalpé 1986), are also included as ericoid fungal species. It is now known (Egger and Sigler 1993; Hambleton and Currah 1997) that one of the most common ericoid fungal species, *Hymenoscyphus ericae*, is the teleomorph of the anamorph fungus, *Scytalidium vaccinii*, both known to form ericoid mycorrhizas (Xiao and Berch 1996). Isolates originally identified as *O. griseum*, based on morphological characters, are now classified as *O. maius*, based on molecular data (Hambleton et al. 1998). In addition to these, many fungal isolates obtained from a range of ericaceous species remain to be identified. Molecular techniques make it possible to compare these isolates with known ericoid fungal species (Perotto et al. 1996; Berch et al. 2002) and, in some cases, identities are confirmed. Molecular methods also show that the root system of any one ericaceous plant may contain a diverse population of mycorrhizal fungi (Perotto et al. 1996; Berch et al. 2002).

Because of the difficulties in identifying fungi isolated from ericaceous plants based on only morphological characteristics, molecular methods have been used to compare isolates from members of the Epacridaceae in Australia with isolates from the Ericaceae in the Northern Hemisphere (McLean et al. 1999). Some epacrid isolates group with *H. ericae*, others with *Oidiodendron,* and others with unknown isolates from species in the Ericaceae (Berch et al. 2002).

Box 14: Ericoid mycorrhizal fungi access various sources of nitrogen

Ericoid mycorrhizal fungi assist a broad spectrum of plants in the large order Ericales in the acquisition of nitrogen. Plant habitat varies considerably but most soils are characterized as being low in nutrients, particularly available nitrogen (Read 1996). Many fungi, including ericoid mycorrhizal fungi, are able to assimilate both ammonium and nitrate obtained from the soil solution (Read 1996). However, other important sources of nitrogen have been documented. *In vitro* experiments have shown that a number of amino acids (Bajwa and Read 1986), peptides and proteins (Bajwa and Read 1985; Bajwa et al. 1985) can be utilized by ericoid fungi as sources of nitrogen. *Hymenoscyphus ericae*, the most common fungal endophyte in the Ericales, produces an extracellular proteinase with a very low pH optimum (pH 2.2). Similar low pH levels may be encountered by this fungus in association with the fine hair roots of ericoid species that occur in the litter layer of acidic heathlands (Leake and Read 1989). Plant, fungal, and animal proteins therefore become potential sources of nitrogen for ericoid fungi and their plant hosts. Nitrogen from proteins that are complexed with polyphenolic compounds can also be accessed by these fungi (Bending and Read 1996); in addition, soluble phenolics and lignin can be degraded by *Hymenoscyphus ericae* (Bending and Read 1997).

Of particular interest, ericoid mycorrhizal fungi are able to use the polymer, chitin, as their sole source of nitrogen (Leake and Read 1990). An important source of chitin for these fungi would be the large biomass of hyphal walls of necrotic mycorrhizal and other fungi as well as insect remains in these soils.

There is convincing experimental evidence that species in the Ericaceae show increased growth when colonized with ericoid mycorrhizal fungi (see Read 1978, 1996); there is less evidence for this in the Epacridaceae (Bell and Pate 1996).

Chapter 5. Arbutoid mycorrhizas

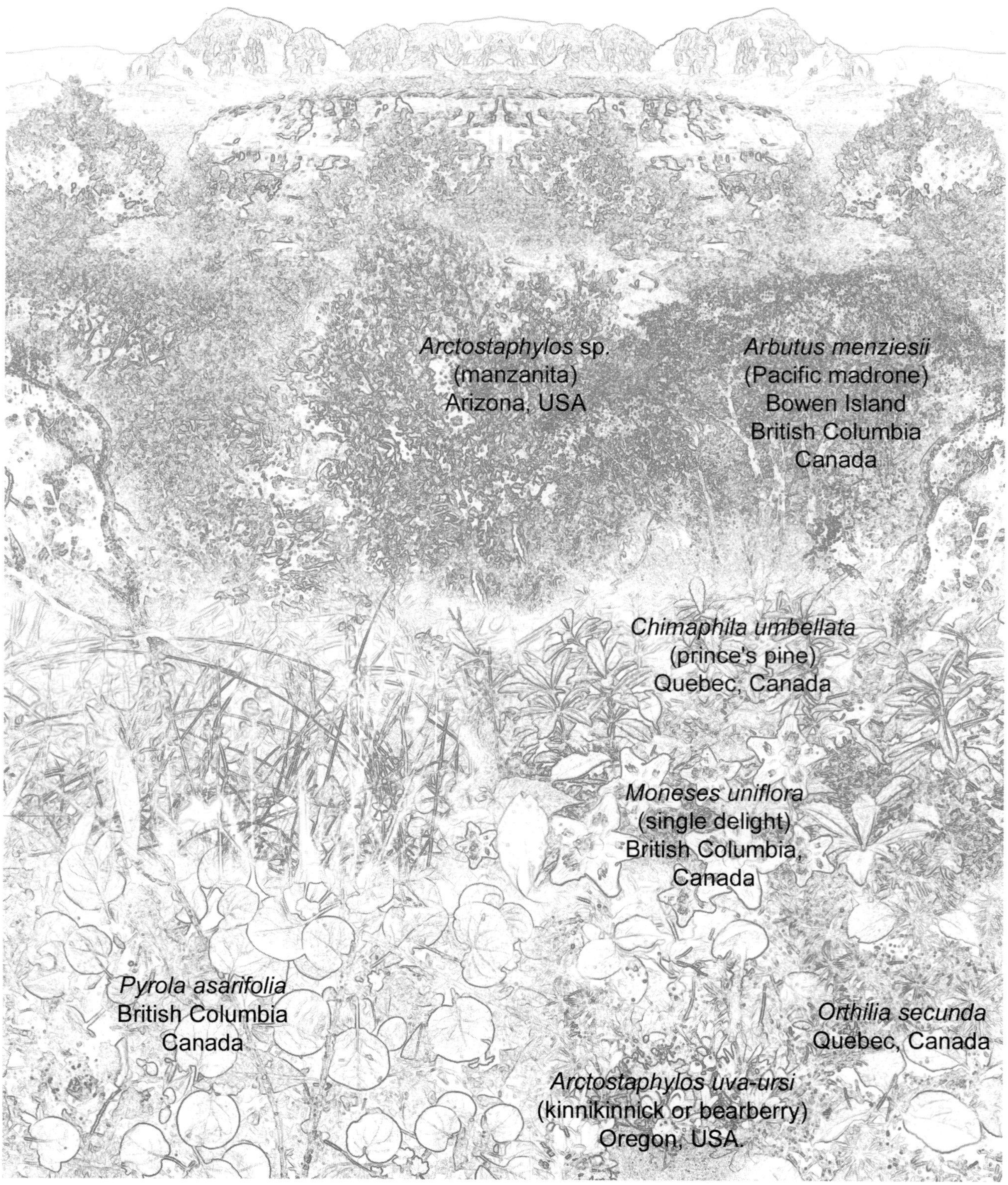
Arctostaphylos sp.
(manzanita)
Arizona, USA
Arbutus menziesii
(Pacific madrone)
Bowen Island
British Columbia
Canada
Chimaphila umbellata
(prince's pine)
Quebec, Canada
Moneses uniflora
(single delight)
British Columbia,
Canada
Pyrola asarifolia
British Columbia
Canada
Orthilia secunda
Quebec, Canada
Arctostaphylos uva-ursi
(kinnikinnick or bearberry)
Oregon, USA.

Chapter 5. Arbutoid mycorrhizas

A. Introduction

1. Definition

Although the plant genera exhibiting arbutoid mycorrhizas are found in two families in the order Ericales, the structural features of the mycorrhizas and the fungal species involved are distinct from those of other members of this order. In certain aspects, they resemble ectomycorrhizas and ectendomycorrhizas and some researchers have included them under the category of ectendomycorrhiza; however, because of structural differences and the fungi involved in the association, it is best to consider them as a distinct category. Structurally, arbutoid mycorrhizas have a mantle, a Hartig net, and intracellular hyphae forming hyphal complexes, the latter two being confined to the epidermis (Figure 179).

2. Plant species involved

Two genera in the Ericaceae (*Arbutus* and *Arctostaphylos*) and several genera in the Pyrolaceae (including *Pyrola*) form typical arbutoid mycorrhizas. Although there are 50 species in the genus *Arctostaphylos*, few have been examined for the presence of mycorrhizas. The two genera *Arbutus* and *Pyrola* have fewer species than the genus *Arctostaphylos*, but again a limited number has been studied in terms of their mycorrhizal associations. Little is known about the mycorrhizal status of other genera in the Pyrolaceae (e.g., *Ramischia, Moneses, Orthilia and Chimaphila*). Although the commercial importance of these genera is mostly as ornamentals, several species form important components of terrestrial ecosystems.

3. Fungal species involved

A number of fungal species that form typical ectomycorrhizas with both gymnosperm and angiosperm tree species also colonize roots of *Arbutus* and *Arctostaphylos* to form arbutoid mycorrhizas under laboratory conditions (Molina and Trappe 1982). Most of these are broad host range fungi, meaning that they show little specificity as to the hosts that they colonize. Narrow host range fungi such as *Alpova diplophloeus*, known to only associate with *Alnus* species, do not form mycorrhizas with either *Arbutus* or *Arctostaphylos*. Roots of *Arbutus menziesii* and *Arctostaphylos manzanita* collected from the field in Northern California were colonized by various species of ectomycorrhizal fungi but formed typical arbutoid mycorrhizas (Acsai and Largent 1983*a,b*). Studies reported in the literature suggest that genera of mycorrhizal fungi are mostly non-specific towards *Arbutus* and *Arctostaphylos* as host plants.

A field experiment in the chaparral on the central coast of California, showed that *Pseudotsuga menziesii* seedlings established only in patches of *Arctostaphylos* and not in patches of *Adenostoma* (an arbuscular mycorrhiza host); identification of ectomycorrhizal fungal species by the presence of sporocarps and by molecular methods led to the conclusion that ectomycorrhizal fungi associated with *Arctostaphylos* contributed to the success of *Pseudotsuga* seedling establishment (Horton et al. 1999).

Most information concerning the fungi forming mycorrhizas with *Pyrola* is based on the hyphal characteristics from roots processed for microscopy. Hyphae with either dolipore septa or simple septa with Woronin bodies have been described, indicating the presence of basidiomycete and ascomycete fungi, respectively. More research, particularly using molecular methods, is needed to determine the range of fungal species able to form arbutoid mycorrhizas.

B. Development and structure

1. Arbutus and Arctostaphylos

Laboratory experiments have shown that the fungal genome may influence the type of root tip branching induced in *Arbutus menziesii* (Pacific madrone) (Massicotte et al. 1993) and *Arctostaphylos uva-ursi* (bearberry) (Zak 1976). In *A. menziesii*, the fungus *Pisolithus tinctorius* induces lateral root formation that results in complex root clusters (Figures 180, 181), whereas the fungus *Piloderma bicolor* induces only a few laterals and no root clusters (Figure 183). In *A. uva-ursi*, depending on the fungal species, either simple root systems with few unbranched laterals form (Figure 182) or root systems with clusters of lateral roots develop. Field-collected roots of *Arbutus menziesii* and *Arbutus unedo* also show variable root branching patterns, presumably due to differences in fungal species forming the mycorrhizal association.

In both *Arbutus* and *Arctostaphylos*, the mantle, depending on the fungal species, can be either compact (Figures 181, 182) and several cell layers thick (Figures 184–187) or loosely-organized with only a few cell layers (Figure 183). A mantle is present during all stages of root cluster formation (Figures 184–187). Rhizomorphs, when present, extend from the mantle surface into the substrate (Figures 181, 182). *Azospirillum*-like bacteria have been isolated from the mantle of *Arbutus unedo* mycorrhizas and transmission electron microscopy has shown that they can occur within the mantle as well as along external hyphae (Filippi et al. 1995).

Hartig net formation occurs only around epidermal cells resulting in a paraepidermal Hartig net (Figures 188, 189). Hartig net hyphae may branch, penetrate the wall of adjacent epidermal cells, and develop further as branched intracellular hyphae (Figure 189). The confinement of Hartig net hyphae to the epidermis may be the result of the suberization of the walls of the outer layer of cortical cells (Figure 190). Intracellular hyphae frequently occupy most of the epidermal cell volume; they are separated from the epidermal cell cytoplasm by the elaboration of a host-derived plasma membrane and an interfacial matrix of undetermined composition (Fusconi and Bonfante-Fasolo 1984). Colonized epidermal cells retain their cytoplasm as well as the usual organelles of living plant cells. The inner mantle, Hartig net and intracellular hyphae all contain abundant mitochondria, endoplasmic reticulum, and ribosomes. These hyphae may also store glycogen and polyphosphate (Ling-Lee et al. 1975). Intracellular hyphae appear to degenerate in older root systems.

2. Pyrola

The root systems of several *Pyrola* species have been illustrated by Robertson and Robertson (1985); some are sparsely branched while others have many laterals, often forming clusters. One of the species illustrated (*P. secunda*) has been renamed *Orthilia secunda*. The role of the fungus in the branching patterns has not been determined. The *Pyrola* mycorrhizas described to date have loosely-organized, thin mantles, a paraepidermal Hartig net that is usually one cell wide, and intracellular hyphae that may fill most of the volume of epidermal cells (Figures 191, 193). The intracellular hyphae are separated from the epidermal cell cytoplasm by host-derived plasma membrane and an interfacial matrix. The nature of this matrix has not been determined. Mantle hyphae and Hartig net hyphae are highly branched and possess numerous septa (Figures 192*a,b*).

C. Functions

Many arbutoid hosts are important components of global ecosystems and may help to maintain the diversity of numerous mycorrhizal fungi. Although they represent few genera, their taxonomic diversity is noteworthy. Estimates are in the order of 50 species for *Arctostaphylos*; this genus is native to North and Central America with *A. uva-ursi* being circumpolar (Young and Young 1992). *Arbutus* species (estimated 10–20 species) are native to the Mediterranean region, Canary Islands and North America (Young and Young 1992), with *A. menziesii* being a prominent coastal tree species from southwestern British Columbia, Canada, to southern California, U.S.A. Members in the family Pyrolaceae include such genera as *Pyrola, Chimaphila, Moneses* and *Orthilia* (with approximately 40 species overall, of which *Pyrola* spp. represent about half), and are principally found in cool north temperate and arctic environments, but do occur as far south as Mexico and the West Indies (Heywood 1993). *Arctostaphylos* and *Pyrola* species are widely distributed in several large forest ecosystems across North America including mixedwood, boreal and sand dune habitats. Some arbutoid species (e.g. *A. uva-ursi*) provide important ground cover and appreciable biomass in many northern regions.

The fact that some (although probably not all) arbutoid hosts associate with mycorrhizal fungi that may also form ectomycorrhizal symbioses with other tree species, suggests a strong potential for the development of fungal linkages and nutrient exchange; in addition, arbutoid species may also act as repository (refuge) for mycorrhizal

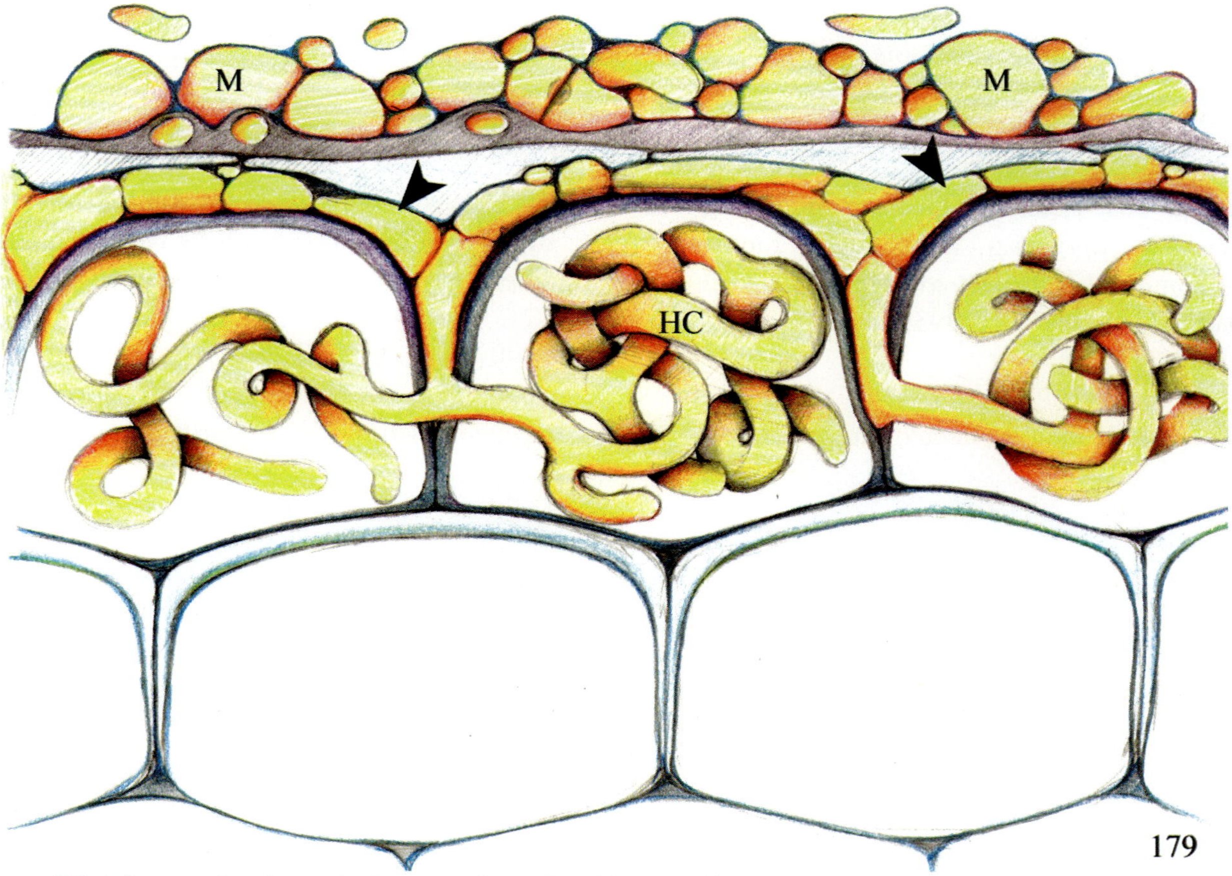

Figure 179. Diagram showing main features of an arbutoid mycorrhiza. A mantle (M), paraepidermal Hartig net (arrowheads), and intracellular hyphal complexes (HC) are present.

fungi, following tree harvest. Exceptions to this "broad" host specificity has been suggested for some fungi, such as certain *Leccinum* species, which may only associate with *Arbutus* or *Arctostaphylos* (Thiers 1975; Acsai and Largent 1983*a,b*; Molina et al. 1992).

Horton et al. (1999) suggest that fungi symbiotic with *Arctostaphylos glandulosa* facilitate the survival and growth of Douglas-fir seedlings growing in the central coast area of California. When grown in *A. glandulosa* patches, 1-year-old Douglas-fir seedlings shared 17 species of fungi with *Arctostaphylos*, allowing for possible fungal linkages and facilitating seedling establishment. A field study by Hagerman et al. (2001) documented that bearberry (*Arctostaphylos uva-ursi*) growing on sites 3 years after harvest in southern British Columbia had 10 morphotypes (out of 17) in common with Douglas-fir seedlings on the same sites.

With this important mycorrhizal group, we are just beginning to explore the importance and magnitude of fungal linkages, as well as refugia for mycorrhizal fungi, and their impact on maintaining species diversity, especially following ecosystem disturbance.

Figure 180. Clusters of roots of *Arbutus menziesii* colonized by *Pisolithus tinctorius*. Numerous extraradical hyphae (arrowheads) are present.

Figure 181. Scanning electron micrograph of root clusters of *Arbutus menziesii* colonized by *Pisolithus tinctorius*. Each lateral root is covered by a mantle (M) of interwoven hyphae. Rhizomorphs (arrowhead) are present in the extraradical mycelium. From Massicotte et al. Mycorrhiza, **3**: 1-11 (1993).

Figure 182. Field collected cluster of roots of *Arctostaphylos uva-ursi* with extraradical hyphae (arrowhead).

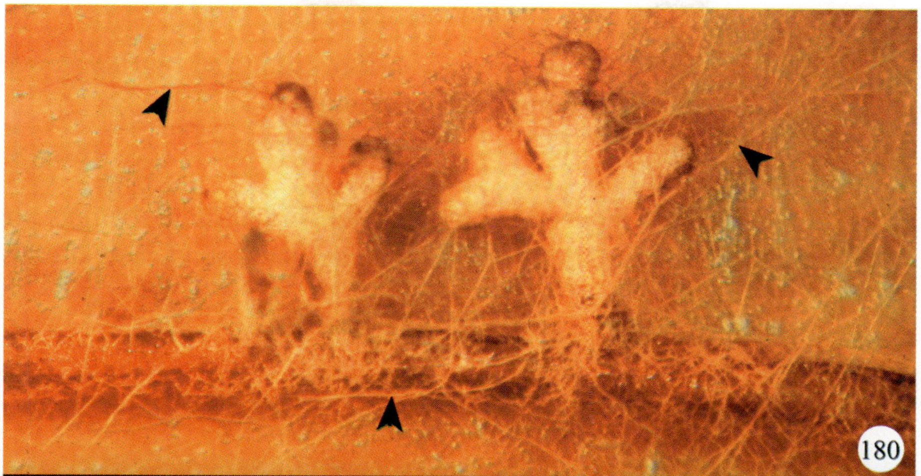
180

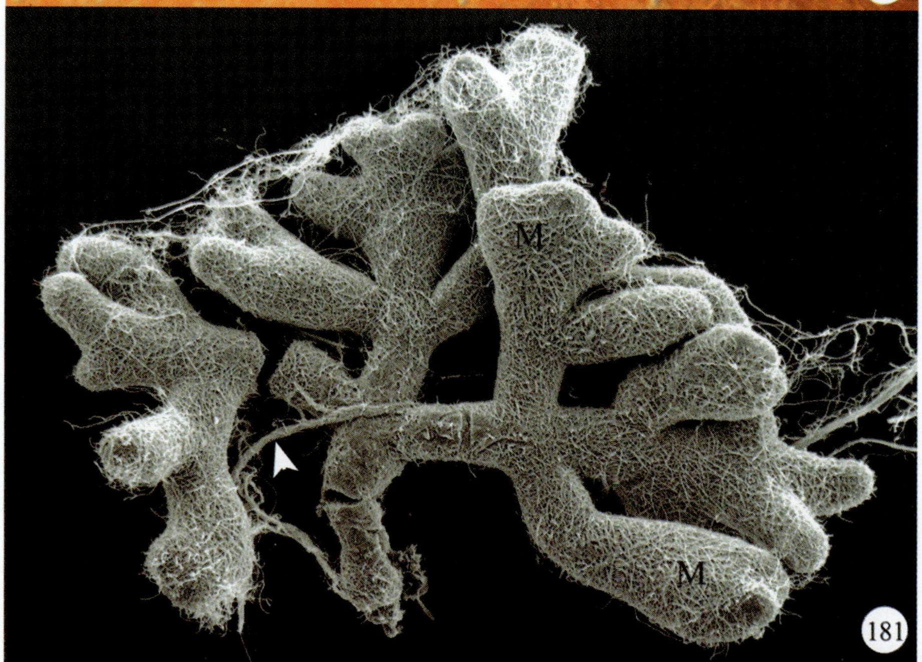
M
M
181

182

Figure 183. Longitudinal section of *Arbutus menziesii* root colonized by *Piloderma bicolor*. The root is unbranched and has a very thin mantle (arrowheads).

Figure 184–187. A developmental sequence of root cluster formation in *Arbutus menziesii* colonized by *Pisolithus tinctorius*. A mantle (arrowheads) is present around each lateral.

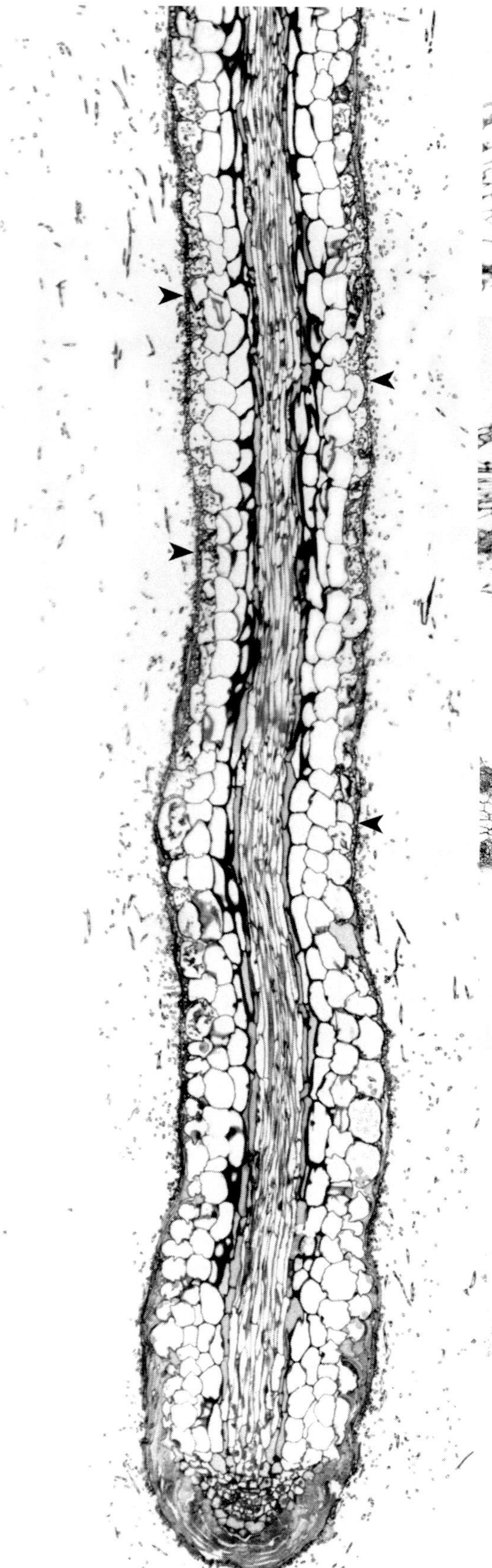

183

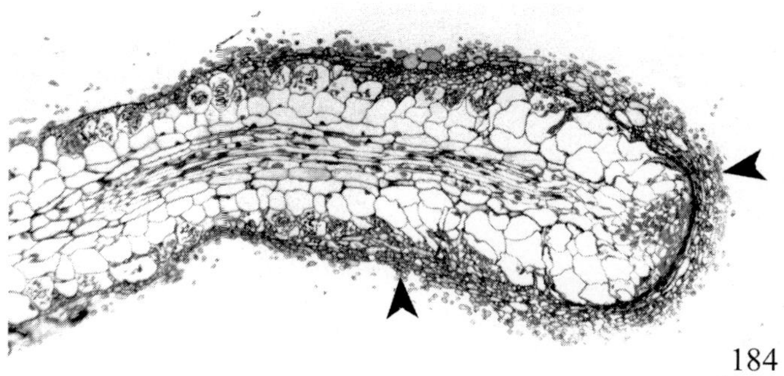

184

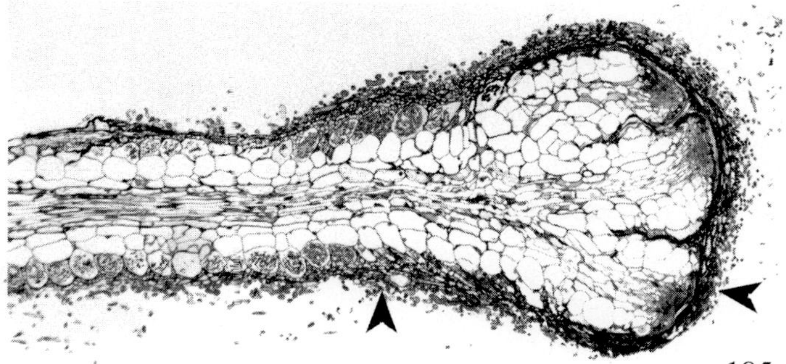

185

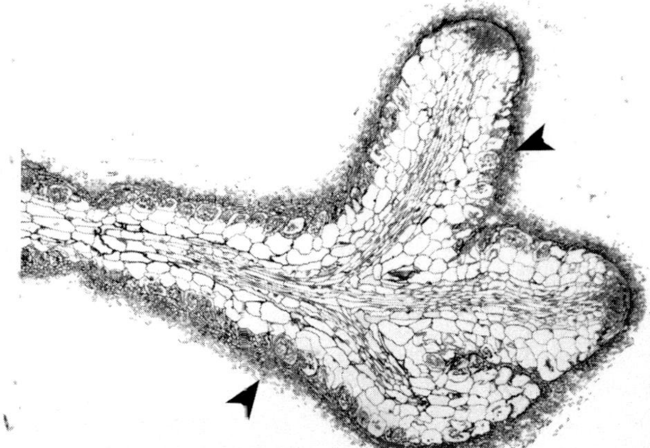

186

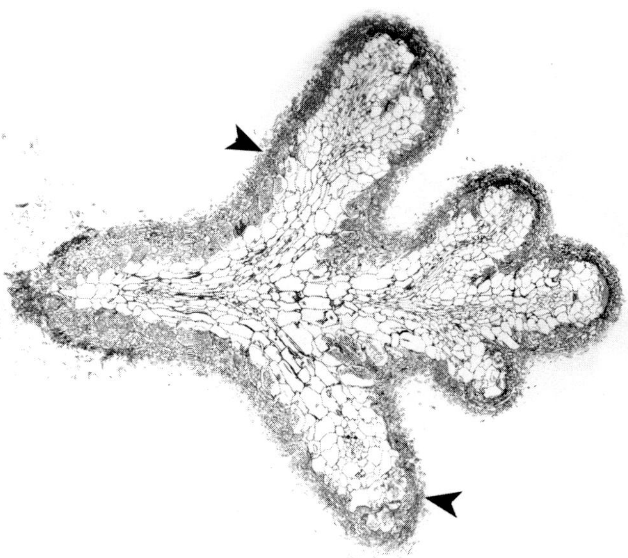

187

Figures 188, 189. Transverse sections of *Arbutus menziesii* roots colonized by *Piloderma bicolor* showing the mantle (M), the restriction of the Hartig net (arrowheads) to the epidermis (paraepidermal), intracellular hyphae (arrows), and the penetration of an epidermal cell by a Hartig net hypha (double arrowhead).

Figure 190. Transverse section of an *Arbutus menziesii* root showing an outer fluorescent layer (the exodermis, arrow), and an inner fluorescent layer (the endodermis, arrowhead). The fluorescence in both cell layers is due to suberin deposition in the walls.

Figure 191. Section of *Pyrola* mycorrhiza with intracellular hyphae filling an epidermal cell (e). Cortex (c), mantle (m), Hartig net (arrowheads).

Figures 192, 193. Confocal laser scanning microscopy of *Pyrola* mycorrhizas.

Figures 192a,b. Sections through the Hartig net showing branching patterns of hyphae.

Figure 193. Hartig net (arrowheads) and intracellular hyphae (arrow).

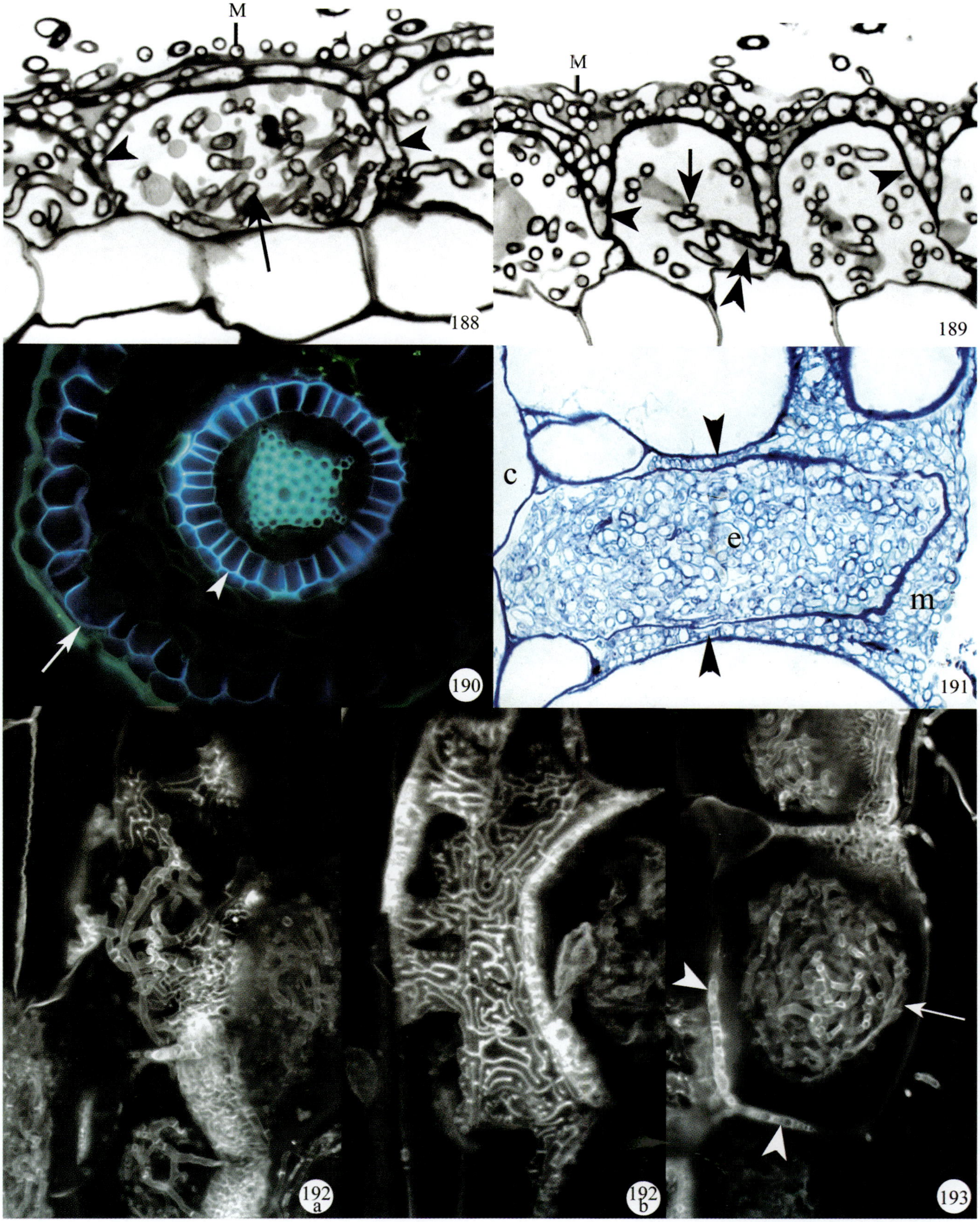
M
188
M
189
190
c
e
m
191
192
a
192
b
193

Chapter 6. Monotropoid mycorrhizas

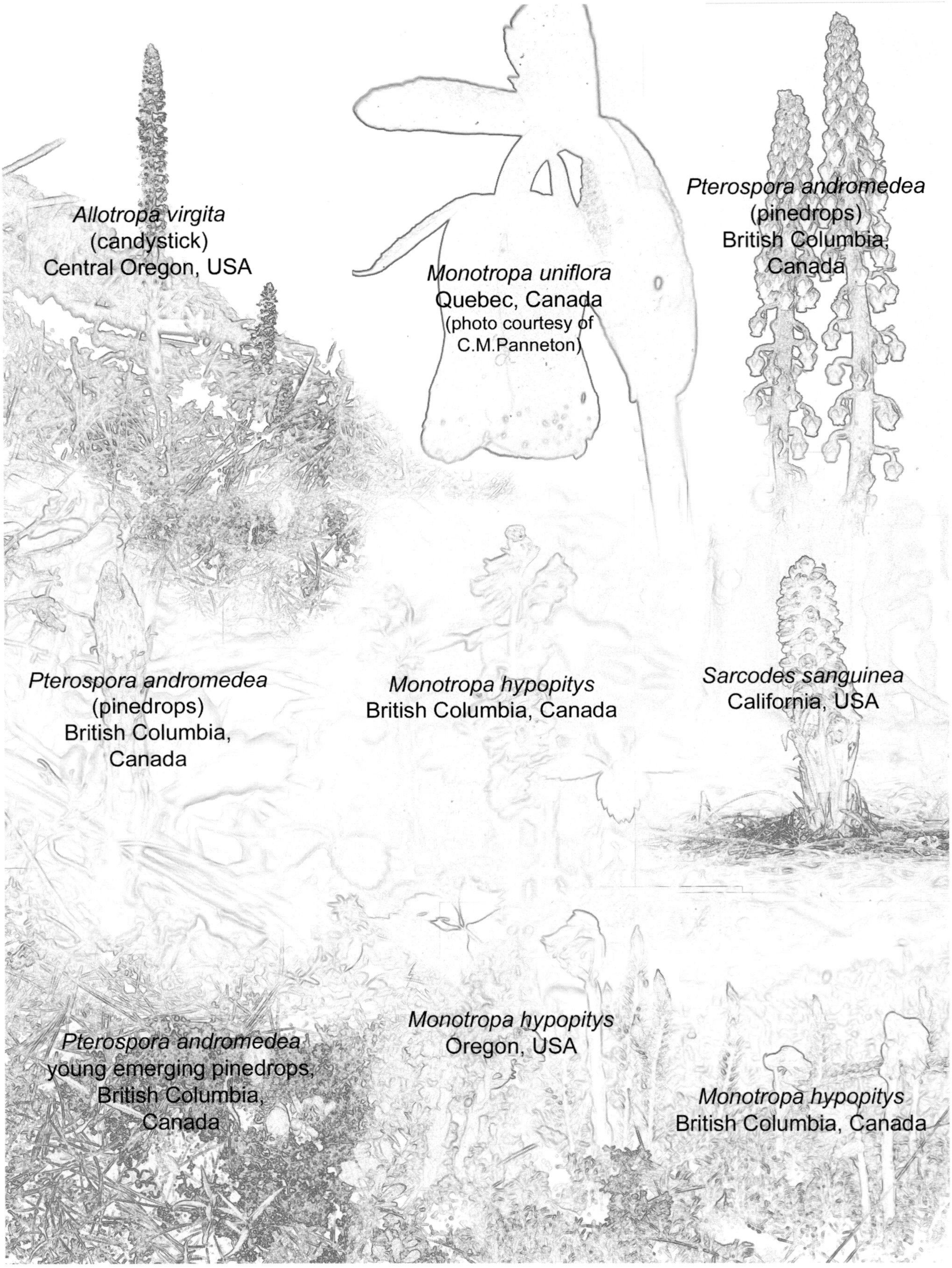
Allotropa virgita
(candystick)
Central Oregon, USA
Monotropa uniflora
Quebec, Canada
(photo courtesy of
C.M.Panneton)
Pterospora andromedea
(pinedrops)
British Columbia,
Canada
Pterospora andromedea
(pinedrops)
British Columbia,
Canada
Monotropa hypopitys
British Columbia, Canada
Sarcodes sanguinea
California, USA
Pterospora andromedea
young emerging pinedrops,
British Columbia,
Canada
Monotropa hypopitys
Oregon, USA
Monotropa hypopitys
British Columbia, Canada

Chapter 6. Monotropoid mycorrhizas

A. Introduction

1. Definition

Although the plant species that form this category of mycorrhiza also belong to the large order Ericales, structural features of the mycorrhiza separate it from ericoid and arbutoid mycorrhizas (Duddridge and Read 1982). Monotropoid mycorrhizas, in common with arbutoid mycorrhizas, share features typical of ectomycorrhizas and ectendomycorhizas. Monotropoid mycorrhizas have a mantle which is sometimes very thick, and a Hartig net confined to the epidermis (paraepidermal). They also possess a unique feature, the invasion of epidermal cells by short hyphae originating from the Hartig net or inner mantle. These structures, referred to as fungal pegs (Figures 194, 195), form along the outer tangential wall of epidermal cells in *Monotropa* species (Figure 194) but at the base of the radial wall of epidermal cells in *Pterospora* (Figure 195) and *Sarcodes sanguinea* (Robertson and Robertson 1982). Host cells respond by depositing additional cell wall material, in finger-like projections, around each peg (Figures 194, 195).

A second feature of monotropoid mycorrhizas is that plants forming this type of mycorrhiza are all achlorophyllous, heterotrophic (nonphotosynthetic) species; they depend on symbiotic fungal associations that act as linkages to neighbouring autotrophic (photosynthetic) trees or shrubs for their carbon acquisition. In an extensive review of heterotrophic plants and their fungal associations, Leake (1994) suggests the term myco-heterotrophic best describes these associations. Others refer to these plants as epiparasites, because they indirectly "parasitize" surrounding trees. It is generally accepted that the term saprophyte should be avoided.

2. Plant species involved

Depending on the taxonomic system used, all genera are found either in the family Monotropaceae or in the clade Monotropoideae, in the family Ericaceae. Regardless, ten genera (*Allotropa, Cheilotheca, Hemitomes, Monotropa, Monotropantham, Monotropsis, Pityopus, Pleuricospora, Pterospora, Sarcodes*) are presently recognized (Leake 1994), all but three of these genera being monotypic (i.e., with only one species). The Asian genus *Cheilotheca* has four recognized species, and the genera *Monotropa* and *Pleuricospora* each have two species. The distribution of the group is primarily north temperate with the largest number of species in western North America. Most of the research, in terms of mycorrhizal associations, has been with the genera *Monotropa* (Indian pipe, Figure 196), *Sarcodes* (the snow plant, Figure 197), and *Pterospora* (pinedrops, Figure 198).

3. Fungal species involved

Over the years, a number of studies have documented the fungi possibly associating with members of the Monotropoideae; with the advancement of molecular approaches, the identities of some fungi have been confirmed (Bidartondo and Bruns 2001). The plant genera examined are associated with members of five families of basidiomycetes, fungi also known to form ectomycorrhizas with numerous tree species. The interesting trend is that each plant genus or species, where there is more than one species in the genus, has become highly specialized in terms of its fungal associates. For example, *Monotropa hypopitys* appears to form monotropoid mycorrhizas only with species in the fungal genus *Tricholoma*, whereas *Monotropa uniflora* forms mycorrhizas with *Russula* species and other members in the family Russulaceae (Bidartondo and Bruns 2001; Young et al. 2002). This specificity is true regardless of the geographical distribution of the plant species. The monotypic plant species, *Sarcodes sanguinea* and *Pterospora andromedea*, shown to be closely related phylogenetically, form mycorrhizas with fungal species in the genus *Rhizopogon*, with each plant species showing preference for particular *Rhizopogon* species.

Bruns and Read (2000) isolated *Rhizopogon* species from mature plants of both *S. sanguinea* and *P. andromedea* and used these, as well as a number of other ectomycorrhizal fungal species, in germination trials with both plant species. Of

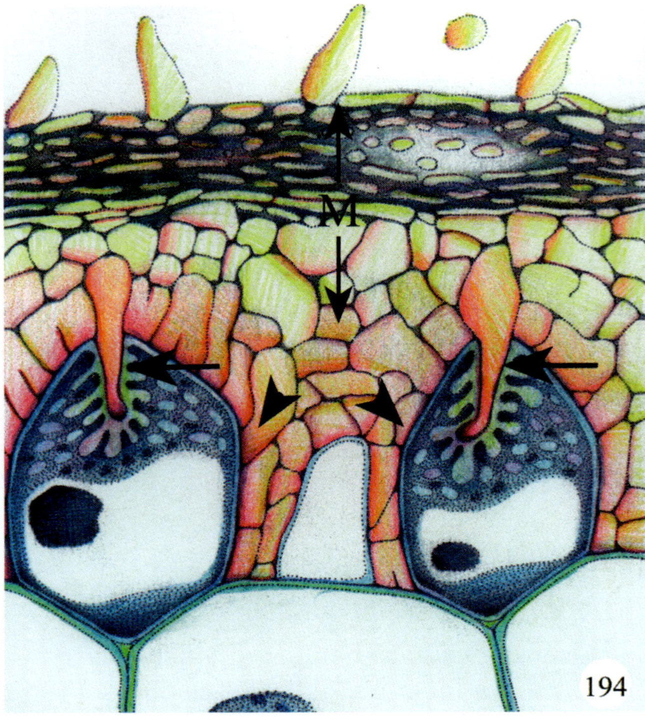

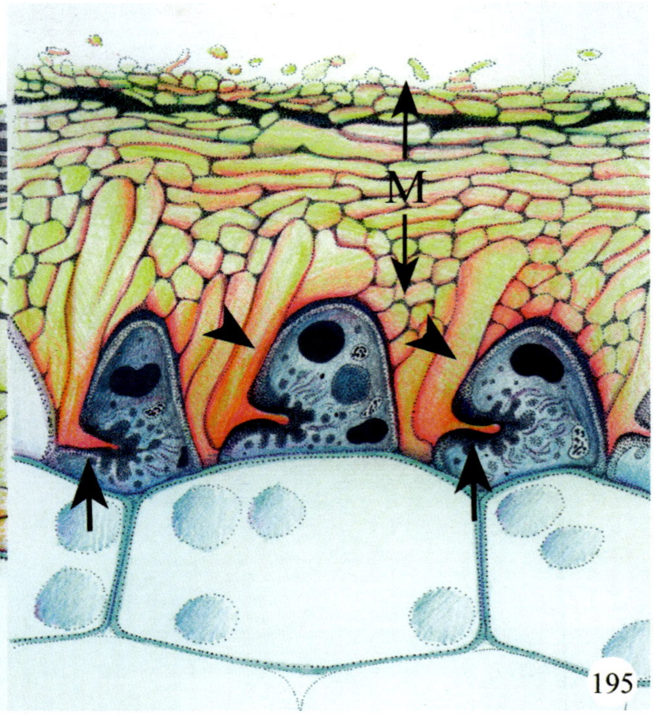

Figure 194. Diagram of main features of a monotropoid mycorrhiza as found in a *Monotropa* sp. A mantle (M), Hartig net (arrowheads) and fungal pegs (arrows) develop. Each peg forms from an inner mantle hypha that enters the cell through the outer tangential wall. Each peg is surrounded by finger-like projections of host-derived wall material.

Figure 195. Diagram of a monotropoid mycorrhiza as found in the genus *Pterospora*. A mantle (M), Hartig net (arrowheads), and fungal pegs surrounded by host-derived wall material (arrows). In this genus, a hypha penetrates the epidermal cell along the base of a radial wall.

the fungi tested, only *Rhizopogon* isolates were effective in promoting seed germination; isolates originating from either plant species were effective in stimulating germination in both plant species. Seeds without a *Rhizopogon* isolate did not germinate. This study shows the importance of the presence of the appropriate fungus for seed germination in these myco-heterotrophic species.

B. Development and structure

The most detailed observations of the development and structure of monotropoid mycorrhizas have been obtained from studies on the genus *Monotropa*. In *M. uniflora* (Figure 199), and *M. hypopitys*, clusters of roots form, each root becoming ensheathed with fungal hyphae resulting in a well-developed mantle (Figure 200). Details of the mantle can best be seen using scanning electron microscopy (Figures 201–203). Frequently, large calcium oxalate crystals are deposited among the outer mantle hyphae (Figure 204) and along the surface of hyphae within the mantle. Cystidia may be present, depending on the fungal species (Figures 203, 205). Rhizomorphs are sometimes present (Figures 206, 207), and have been traced *in vivo* from the surface of *M. hypopitys* mycorrhizas to the roots of neighbouring pine trees (Duddridge and Read 1982).

The mantle is usually multi-layered (Figures 208, 209) with a compact inner mantle (Figure 210). A paraepidermal Hartig net develops (Figures 208, 209) and hyphae originating from this or from inner mantle hyphae penetrate into epidermal cells to form fungal pegs (Figures 210, 211). These fungal pegs, one per epidermal cell, are encased by wall material synthesized by the host plant. The wall is laid down unevenly so that each peg has wall projections enveloped by host plasma membrane (Figure 211), a structure similar to that in transfer cells described in many plant tissues. The formation of plasma membrane around these wall ingrowths increases the surface area of this membrane; increased surface area may be important in nutrient exchange at this interface. Mitochondria and profiles of endoplasmic reticulum, often suggestive of increased metabolic activity, are always found in the vicinity of these wall ingrowths. The pegs form along the outer tangential wall of epidermal cells in *Monotropa*. In any cross section of a root, most epidermal cells show the development of these structures. With time, each peg undergoes changes in that the host wall surrounding the tip breaks down; the fungal peg eventually degrades and, concomitantly, changes occur in the surrounding plasma membrane to form a sac-like structure. The contents of the fungal peg apparently move into this membranous structure.

In one study of *M. hypopitys*, certain stages in the development of mycorrhizas were documented to correlate with stages in plant development (Duddridge and Read 1982). Mantle and Hartig net formation and the first stages of peg formation in epidermal cells occurred before the shoot emerged above ground. At this time, glycogen was stored in the Hartig net and mantle hyphae. As shoots emerged, the number of fungal pegs increased dramatically and, with time, the number of pegs with degraded tips dominated the population. Following maturation of flowers and during seed set, degeneration of the fungal pegs, Hartig net and mantle hyphae occurred.

Roots of *Pterospora andromedea* also occur in coralloid clusters (Figure 212) with each root forming several laterals (Figure 213). The mantle covers the main root and lateral roots (Figure 214); hyphae of the external mantle are very irregular in shape and small crystals are present on their surface (Figures 215, 216). The mantle is multi-layered (Figures 217, 218) and the Hartig net is paraepidermal (Figures 218, 219). Hyphae from the Hartig net penetrate epidermal cells, usually towards the base of radial walls, to form fungal pegs (Figure 220).

General descriptions of field-collected mycorrhizas of *Sarcodes sanguinea* were included as part of a study on the transport of radioactive compounds from autotrophic trees to this species (Vreeland et al. 1981). *S. sanguinea* mycorrhizal roots developed in clusters or coralloid masses, within which each mycorrhizal root tip formed a mantle and paraepidermal Hartig net. In a more detailed ultrastructural study of this species, as well as *Pterospora andromedea,* Robertson and Robertson (1982) illustrated the coralloid clusters of roots for both species and provided details of the interface between fungus and root cells.

C. Functions

All plant species having monotropoid mycorrhizas are nonphotosynthetic and myco-heterotrophic. Increasing evidence suggests that these specialized plants have evolved a mechanism to obtain photosynthates by forming mycorrhizas with fungi that are also able to associate with neighbouring photosynthetic trees or shrubs. Using radioisotopes, it has been demonstrated that photosynthates (Björkman 1960) can move from trees to *Monotropa hypopitys* plants and that phosphorus (Vreeland et al. 1981) can move from trees to *Sarcodes sanguinea*. The mechanism of transport of nutrients from the mycorrhizal fungus into roots of monotropoid species remains uncertain. The elaborate wall ingrowths and the concomitant development of the plasma membrane around the fungal pegs is circumstantial evidence that this region may be the site of transfer of compounds from the fungus to epidermal cells. The role of the Hartig net in transferring metabolites has not been explored.

Figure 196. Inflorescence of *Monotropa uniflora* (Indian pipe) photographed in Quebec, Canada. Photo courtesy of C.M. Panneton.

Figure 197. Inflorescence of *Sarcodes sanguinea* (the snow plant) growing in California, USA.

Figure 198. Inflorescences of *Pterospora andromedea* (pinedrops) photographed in central British Columbia, Canada.

196

197

198

Figure 199. Large root cluster (arrows) and new shoots (arrowheads) of *Monotropa uniflora*. All root tips appear to be mycorrhizal, probably with a fungal symbiont in the family Russulaceae.

Figure 200. Mycorrhizal root tip of *Monotropa uniflora* showing compact mantle (*) and bristle-like cystidia (arrowheads); these features are common to some fungal members of the Russulaceae. Photo courtesy of B. Young.

Figures 201–205. Scanning electron micrographs of *Monotropa uniflora* mycorrhizas showing features of the mantle.

Figure 201. A main root with several developing lateral roots (arrowheads), all covered by mantle hyphae.

Figure 202. Surface features of a root apex showing abundant cystidia (arrowheads) and crystals (arrows).

Figure 203. Higher magnification of portion of a mycorrhiza showing large numbers of awl-shaped cystidia (arrowheads).

Figure 204. A large crystal of calcium oxalate and fusiform to flask-shaped cystidia (arrowheads), some having small apical knobs.

Figure 205. Sectional view of mycorrhiza showing two types of cystidia, awl-shaped (black arrowheads) and flask-shaped (white arrowheads).

Figures 199-205 from Massicotte et al. Mycorrhiza, **14**. (2004)

Figures 206, 207. Features of rhizomorphs found on *Pterospora* mycorrhizas. Scanning electron microscopy.

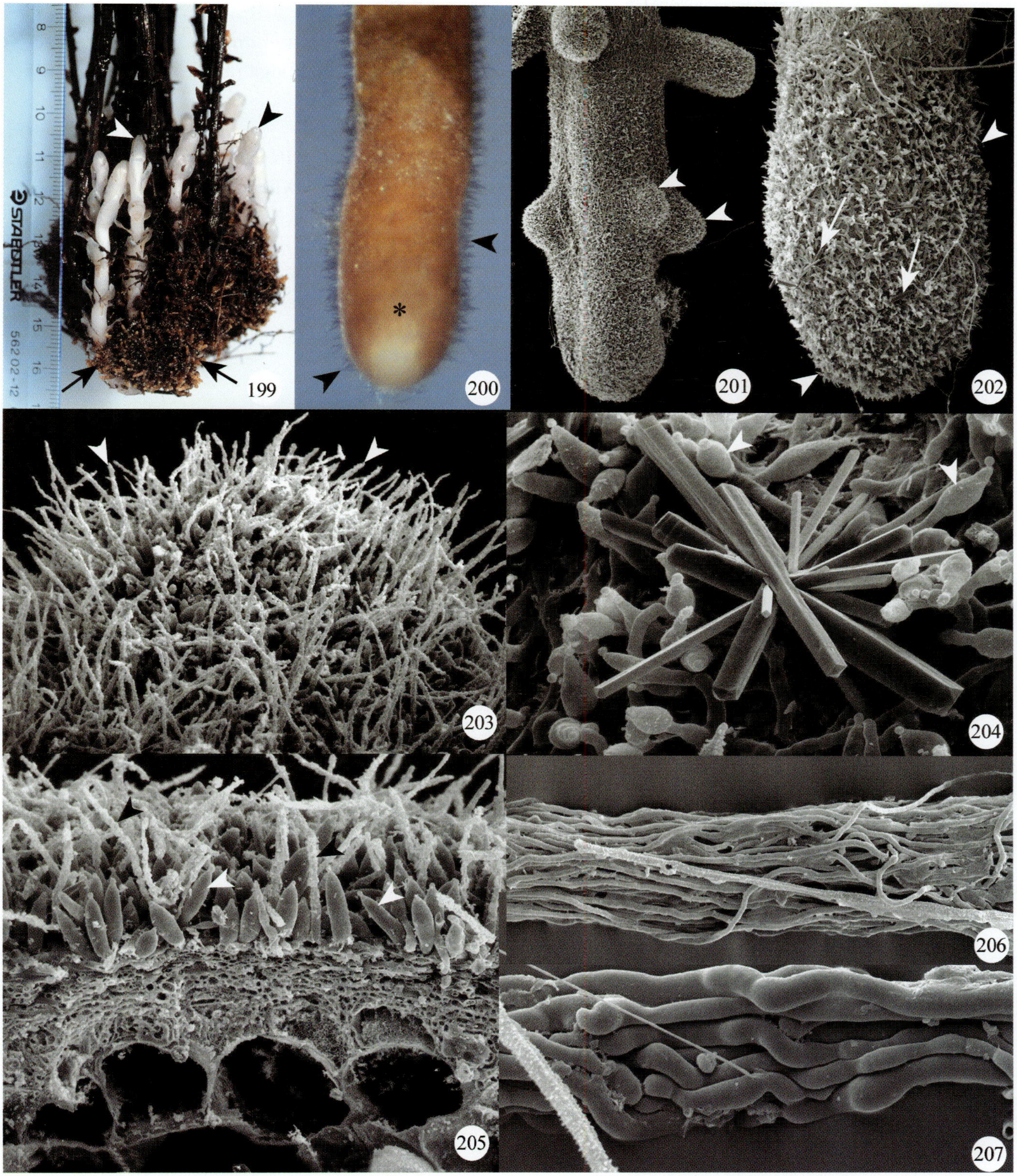
199
200
201
202
203
204
205
206
207

Figures 208–209. Light microscopy of transverse sections of *Monotropa uniflora* mycorrhizas. A thick mantle (M), a paraepidermal Hartig net (arrowheads) and hyphal pegs (arrows) are evident.

Figure 210. Light microscopy of a paradermal section of *Monotropa uniflora* mycorrhiza showing the compact inner mantle (M), and sectional views of hyphal pegs (arrowheads) in epidermal cells.

Figures 208-210 from Massicotte et al. Mycorrhiza, **14**. (2004)

Figure 211. Transmission electron micrograph showing details of a hyphal peg (*). Finger-like depositions of host cell wall material (arrowheads) are evident. Photo from Lutz and Sjolund Am. J. Bot. **60:** 339–345 (1973).

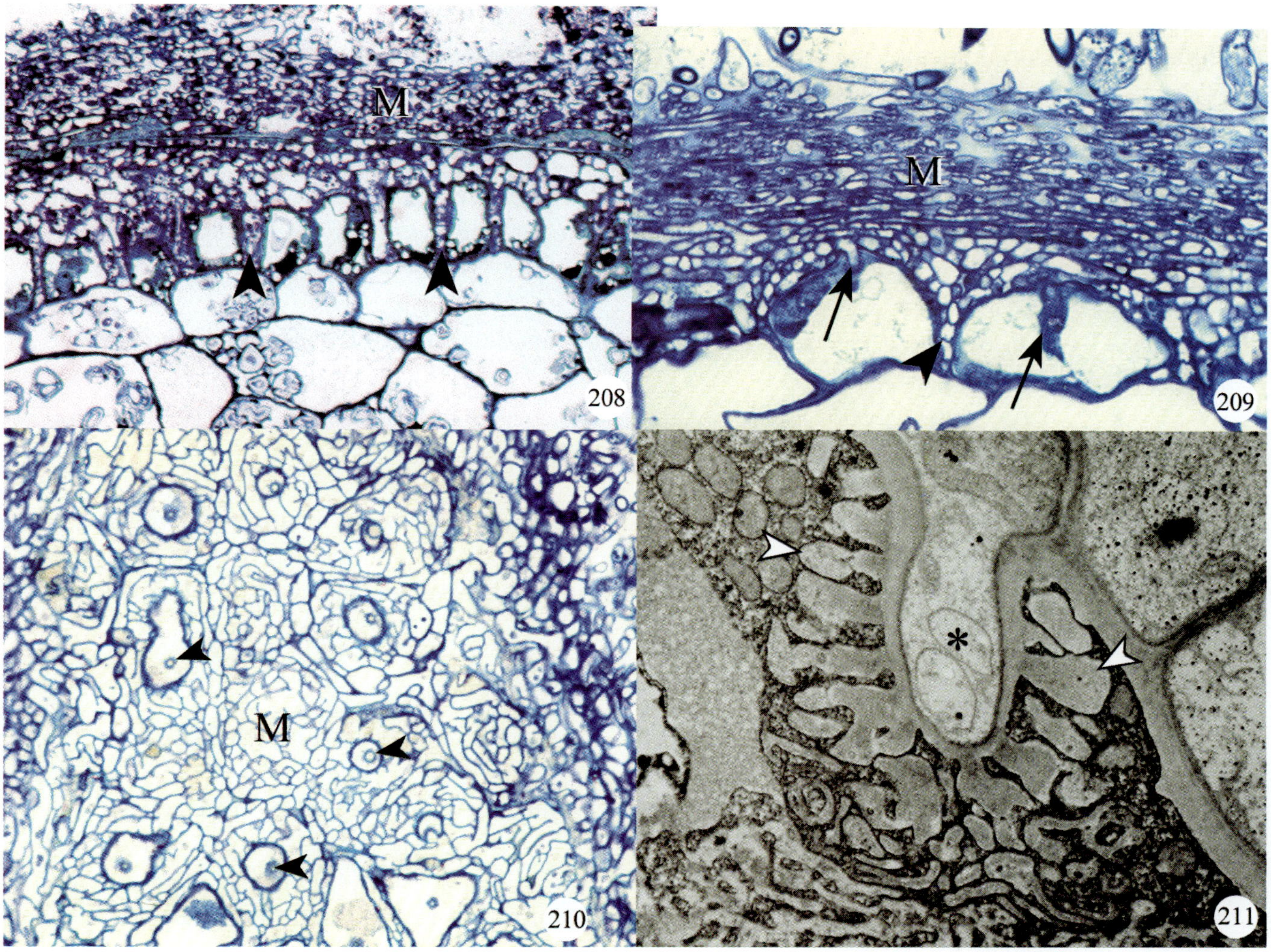
M
208
M
209
M
210
*
211

Figures 212. Coralloid root mass of *Pterospora andromedea* with fungi closely resembling the *Rhizopogon subcaerulescens* group, collected from British Columbia, Canada.

Figure 213. Enlarged view of *Pterospora andromedea* mycorrhiza showing a main root apex and several lateral roots (arrows). A mantle (*) covers all roots. Photo courtesy of J. Catherall.

Figures 214–216. Scanning electron micrographs of mantle surface of *Pterospora andromedea* mycorrhizas.

Figure 214. Mantle surface of main and lateral roots.

Figure 215. Higher magnification of mantle surface showing irregular features of hyphae.

Figure 216. Mantle surface showing irregular hyphae and abundant small crystals (arrowheads) of variable shapes.

Figures 217–220. Light microscopy of sections of *Pterospora andromedea* mycorrhizas.

Figure 217. Portions of two lateral roots showing that they are covered by a mantle (arrowheads).

Figure 218. The thick, layered mantle (*) and paraepidermal Hartig net (arrowheads).

Figure 219. Inner mantle (M), paraepidermal Hartig net (arrowheads), and evidence of hyphal pegs (arrow).

Figure 220. Hyphal pegs (arrowheads) that enter epidermal cells along a radial wall.

Figures 212-220 from Massicotte et al. Mycorrhiza, **14**. (2004)

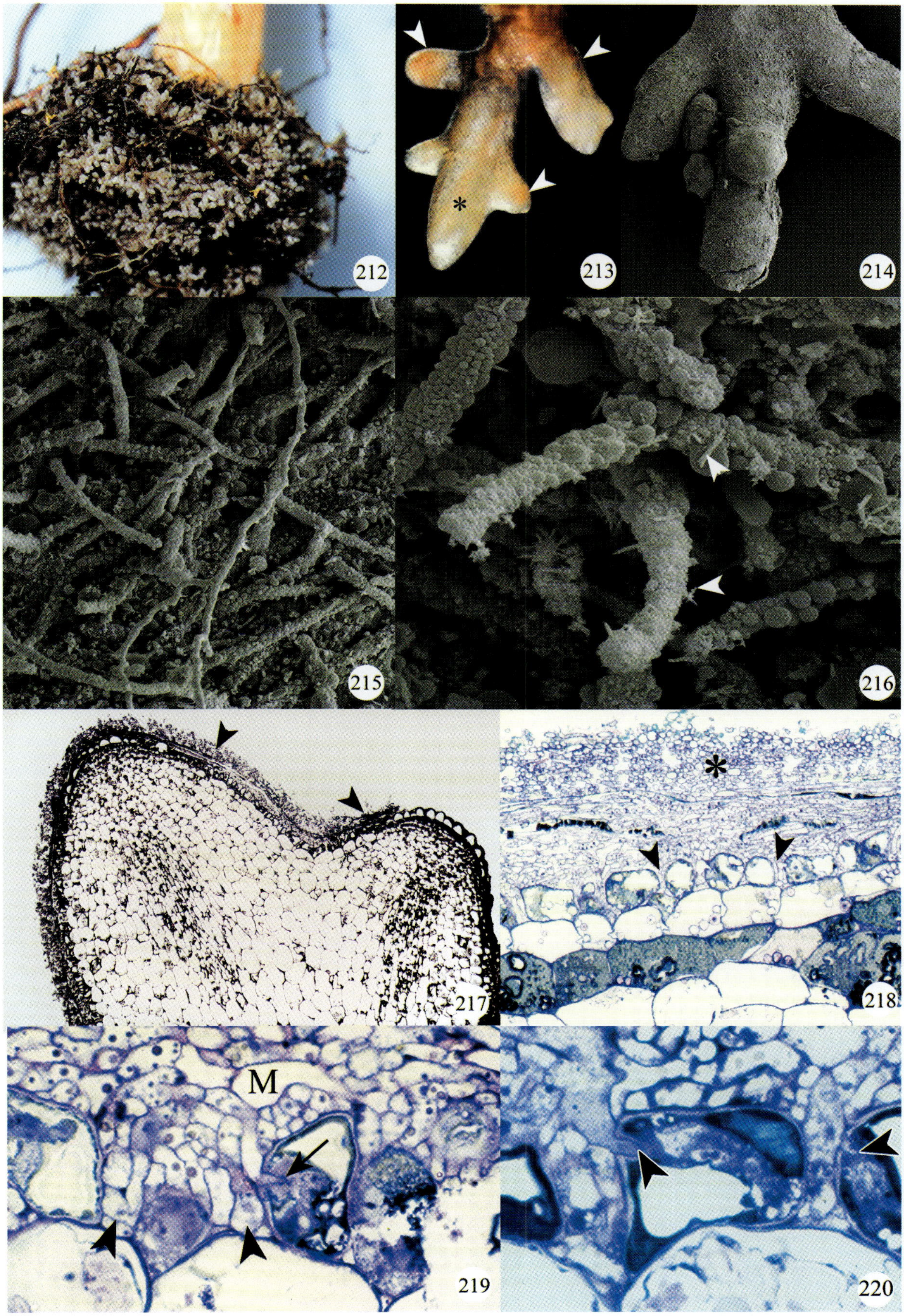
212
213
214
215
216
217
218
M
219
220

Chapter 7. Orchid mycorrhizas

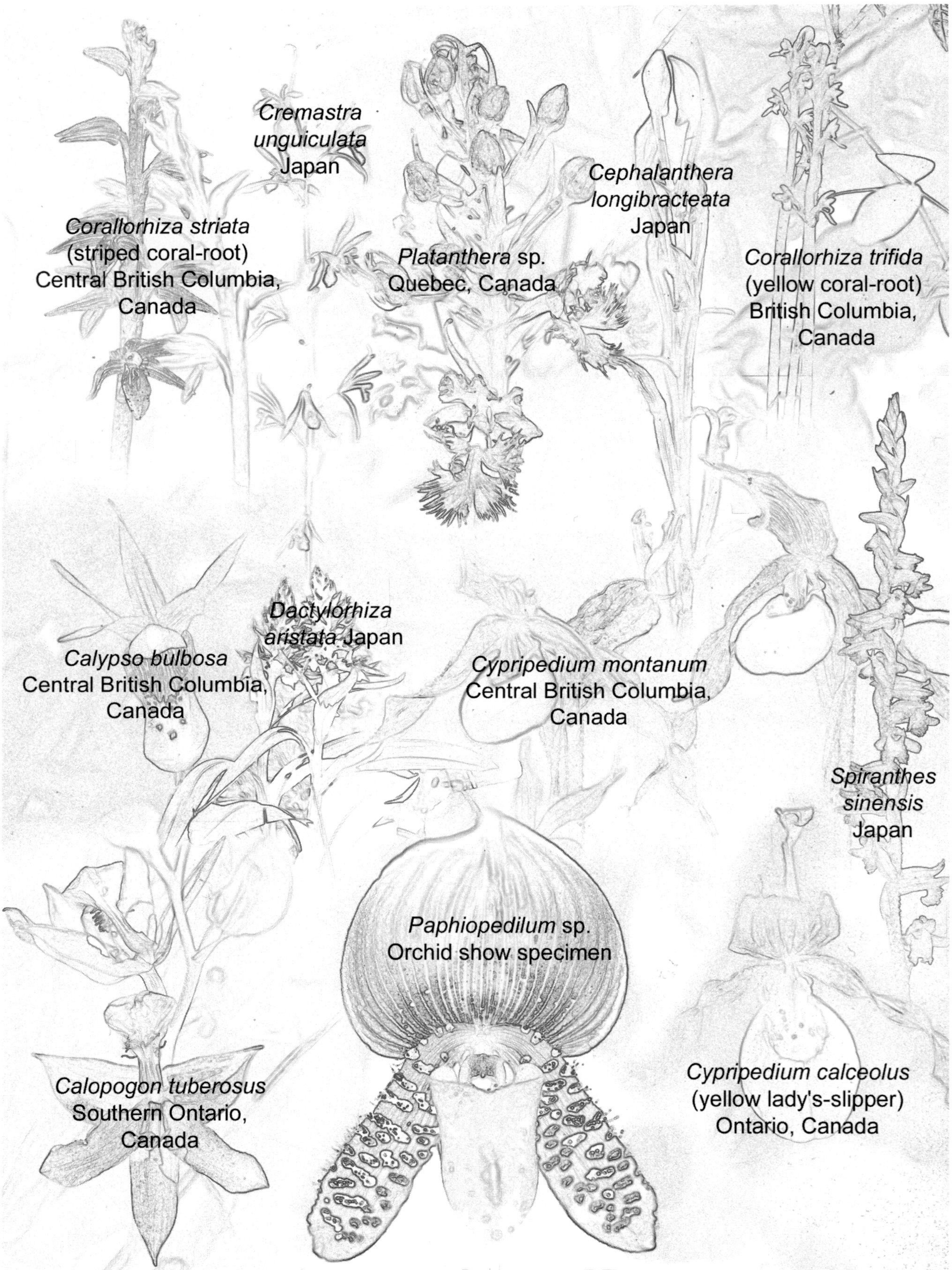

Cremastra unguiculata Japan
Corallorhiza striata (striped coral-root) Central British Columbia, Canada
Platanthera sp. Quebec, Canada
Cephalanthera longibracteata Japan
Corallorhiza trifida (yellow coral-root) British Columbia, Canada
Dactylorhiza aristata Japan
Calypso bulbosa Central British Columbia, Canada
Cypripedium montanum Central British Columbia, Canada
Spiranthes sinensis Japan
Paphiopedilum sp. Orchid show specimen
Calopogon tuberosus Southern Ontario, Canada
Cypripedium calceolus (yellow lady's-slipper) Ontario, Canada

Chapter 7. Orchid mycorrhizas

A. Introduction

1. Definition

Orchid mycorrhizas are unique in that they occur only within the family Orchidaceae, one of the largest families of flowering plants. Although they share some characteristics with other mycorrhizal types (e.g., root cells are colonized by fungi), they differ in that fungal associations are essential for both seed germination and seedling establishment in nature (Rasmussen 1995; Peterson et al. 1998). The main defining characteristic of orchid mycorrhizas is the formation of complex hyphal coils (pelotons) within host plant cells (Figure 221). Orchid mycorrhizas can, therefore, be considered within the broad category of endomycorrhizas. Although it is often presumed that orchid mycorrhizas are mutualistic symbioses, there is no evidence, to date, that the fungus benefits from the association. The suggestion has been made (Smith and Read 1997) that orchids may have evolved a unique relationship with their associated fungi that involves parasitism, in this case by the host plant on the fungus.

2. Plant species involved

The family Orchidaceae, consisting of approximately 450 genera and over 17,000 species, has a worldwide distribution, but with the largest number of species in the tropics. Various life forms occur: terrestrial, epiphytic, lithophytic (plants growing on cliffs or rock faces), with a few species of *Rhizanthella* living entirely below ground until flowering (Dixon 2003). Several orchid species remain achlorophyllous during their entire life cycle and are, therefore, myco-heterotrophic, depending on fungi for carbon compounds derived either from the breakdown of organic matter in the soil or from mycorrhizal linkages with autotrophic plants.

The diversity in floral forms (Figures 222–225) and the specialized pollination mechanisms have attracted considerable attention to the Orchidaceae by scientists, amateur orchid growers, and horticulturalists. Many species are grown commercially for the floriculture industry, and the genus *Vanilla* is grown for its extract from seed pods which is used as a flavouring in the food industry. Many orchid species can now be propagated from seeds *in vitro* by supplying the germinating seeds with a source of simple sugars or providing the appropriate fungus using a medium containing a complex organic source (Figure 226). This is of particular importance since species or hybrids with showy and unusual flowers can be produced in large numbers for the flower market. Also, attempts are being made to reintroduce into their native habitats seedlings of endangered species that have been propagated *in vitro*. A major difficulty with this method is the loss of large numbers of seedlings as they are removed from culture flasks and transplanted to potting soil. Trials are being conducted to inoculate seedlings with appropriate mycorrhizal fungi to overcome this problem at this step in the production process.

3. Fungal species involved

Fungi can be isolated rather easily from roots of orchids and most isolates grow well on various culture media. One method involves removing individual pelotons from colonized root cortical cells and then plating these on sterile media. Hyphae growing out from these pelotons are then subcultured. This method avoids most contamination that could come from other fungi that may be present on the surface or within root tissues. It has been more difficult to isolate fungi from germinating seeds and protocorms. However, it is now possible to place mesh bags containing orchid seeds in soil to act as baits for mycorrhizal fungi (see Rasmussen 1995).

Most of the early descriptions of fungal isolates were based on morphological characteristics of the fungal colonies grown on agar medium (Figures 227–229) in combination with structural characteristics of the hyphae (Currah et al. 1997, Figures 230, 231). Based on this, fungi isolated from roots of orchid species were grouped in the form genus *Rhizoctonia* and many of these were capable of stimulating seed germination of a number of orchid species (Rasmussen 1995). Following intensive culturing of isolates, sexual stages (teleomorphs) have been produced for several isolates and it is now known that there are several anamorphs (asexual stages) with associated teleomorphs that are symbionts with orchids (Currah et al. 1997; Rasmussen 1995). The identification of fungal symbionts in orchid mycorrhizas has also been enhanced recently by the use of molecular methods.

Figure 221. Diagrammatic representation of an orchid root showing the type of colonization typical of orchid mycorrhizas. Hyphae enter the root through the epidermis (arrowheads), and form hyphal coils (pelotons) in cortical cells (arrows). Hyphal coils undergo degradation over time (double arrowheads).

221

Figures 222–225. Various floral forms in the family Orchidaceae.

Figure 222. *Calapogon tuberosus*.

Figure 223. *Cypripedium montanum*.

Figure 224. *Paphiopedilum* sp.

Figure 225. *Oncidium* sp.

222

223

224

225

Some anamorphs that are known to be pathogens on a variety of plant hosts, are able to form symbiotic associations with orchids, at least in sterile culture experiments. Although the systematics of the fungi associated with orchids is still in progress, some facts are known. For example, autotrophic orchid species (those that have chlorophyllous seedlings and adults) associate with different fungal species than myco-heterotrophic species (those that remain achlorophyllous throughout their life cycle). Autotrophic terrestrial orchids form mycorrhizas with basidiomycete anamorph genera such as *Ceratorhiza, Epulorhiza,* and *Moniliopsis* (corresponding teleomorph genera are *Ceratobasidium, Tulasnella* and *Sebacina,* and *Thanatephorus*, respectively) (Currah et al. 1997).

Many of these orchid fungal symbionts produce enzymes such as cellulases and polyphenol oxidases that enable them to break down soil organic matter to simple sugars that may be utilized by both partners in the symbiosis. If necessary these fungi could potentially survive as saprophytes in the soil, and act as viable sources of inoculum for developing orchids.

Myco-heterotrophic orchid genera such as *Corallorhiza*, however, are dependent on fungal symbionts for their nutrition and form mycorrhizal associations with several basidiomycete fungal genera known to form ectomycorrhizas with tree species. Among these, species of *Russula, Thelephora*, and *Tomentella* are important. As described for the plant members of the Monotropoideae, myco-heterotrophic orchid genera are able to form mycorrhizas with fungal species that link them with adjacent trees or shrubs and use these hyphal linkages between their roots and tree roots to gain photosynthates produced by the autotrophic host **(see Box 15)**. Fungal linkages between orchids and tree and (or) shrub species are being reported more frequently and are indicative of the complex ecological relationships that exist among plants and other organisms in nature.

B. Orchid seed germination and protocorm formation

1. Seed and protocorm structure

A detailed account of orchid seed germination in relation to symbiotic fungi can be found in the review by Peterson et al. (1998). All orchid species form numerous minute seeds per capsule and, because of their small size, are referred to as 'dust seeds' (Figure 232). Seed coats display various patterns (Figure 233), some of which may be related to their dispersal by wind. Each seed contains a minute undifferentiated embryo lacking a root and shoot apical meristem (Figure 234). In addition, the embryo cells store lipids and proteins (Figure 235) since endosperm is not present. These unique structural characteristics require that seeds must first become colonized by an appropriate fungal species that provides necessary carbohydrates for further development of the embryo into a structure called a protocorm (see Figures 236–239 for the sequence of events that occur from seed to protocorm to seedling). The protocorm is composed of parenchyma cells, some of which eventually initiate a shoot apical meristem. The first adventitious root primordium is initiated after the first leaf primordium forms from the shoot apical meristem. An unusual feature of orchid seedlings is that a primary root system is lacking so that all roots formed are adventitious in origin. Since the protocorm and the seedling that develops from it often remain below ground for more than one growing season, this phase of the life cycle is myco-heterotrophic.

2. Early colonization events

Most of the information concerning colonization of orchid embryos comes from *in vitro* experiments, primarily because of the difficulty in following the early developmental stages in soil. Recently, the use of small mesh bags containing seeds, and placing these in soil so that seeds can become colonized by fungal hyphae, has led to some information on natural infection processes (see Rasmussen 1995). A hypha of a compatible fungal species contacts an imbibed seed and enters either through the suspensor end of the embryo (Figure 240) or through epidermal hairs (some authors refer to these as rhizoids) that develop from the embryo after imbibition and the splitting of the seed coat (Figure 234). Once inside the embryo, the fungus forms hyphal coils (pelotons) in the embryo cells (Figures 240–242). Most cells, except those from which the shoot apical meristem develops (and in some species, the epidermis), become colonized. The hyphal coil is separated from host cell cytoplasm by a perifungal membrane (Figure 243) that is a modification of the host plasma membrane. Entry of fungal hyphae into embryo and protocorm cells triggers nuclear enlargement within host cells (Figure 244); this event is accompanied by several cycles of DNA synthesis and changes in the host cell cytoskeleton **(see Box 16)**.

Fungal hyphae pass from cell to cell presumably by the production of hydrolytic enzymes that break down localized areas of cell wall. Hyphae that pass through the host cell wall have a very narrow diameter compared to peloton hyphae (Figure 245).

The embryo increases in size through cell divisions and cell enlargement to produce the protocorm (Figures 246, 247). As this happens, the first formed pelotons undergo degradation to form clumps of lysed hyphae within cells (Figure 246). It is now known that these degraded hyphae are isolated from the host cell cytoplasm by wall constituents synthesized by the host cell (Peterson et al. 1996, Figures 248, 249). This may be important in preventing host cytoplasm damage during this lytic process and may allow the cell to become invaded more than once by fungal hyphae.

C. Seedling establishment and mature plants

Once the shoot apical meristem is established, leaves form and root primordia are initiated (Figure 239). Many orchid species remain below ground for more than one season or, in some cases, throughout the vegetative phase of their life cycle. Consequently, they depend on their fungal partner to supply carbohydrates for growth. Eventually, in chlorophyllous orchids, leaves emerge above ground and these are able to synthesize their own sugars. As roots elongate into the soil they are colonized by symbiotic fungi, and cortical cells develop pelotons (Figures 250, 251). It appears that the roots are colonized *de novo* from fungal propagules in the soil, and not by the fungus passing from the protocorm. Thus different fungal species may colonize the root and the protocorm, and more that one fungal species may invade the same root.

There are few studies of the colonization of orchid root cells by mycorrhizal fungal hyphae, but reports indicate that fungi may enter directly through epidermal cells or root hairs. In orchid roots with an exodermis consisting of short and long cells, hyphae enter through the short cells to gain access to the remainder of the cortex. This is analogous to what occurs in roots with an exodermis that are colonized by AM fungi. Aerial roots of epiphytic orchids have a multi-layered epidermis (the velamen) that can become colonized by a variety of organisms (Zelmer 2001; Figures 252–255). These roots also have an exodermis consisting of short and long cells and, as above, mycorrhizal fungal hyphae access the cortex by passing through the short cells (Figure 252).

Pelotons occurring in root cells are degraded and isolated from the host cytoplasm, following which a cortical cell can be reinvaded by the fungus (Figure 251). Terrestrial orchid species may develop a variety of underground organs including roots, root tubers (roots specialized for storage of starch), and rhizomes. There are reports of all of these structures being associated with mycorrhizal fungi, but the highest levels of colonization are generally found in the roots. Field-collected roots that appear white are usually filled with starch and colonization levels of mycorrhizal fungi are often low. Older roots that appear dark in colour due to the accumulation of phenolics in epidermal cells, are usually heavily colonized by mycorrhizal fungi.

D. Functioning of orchid mycorrhizas

1. Seed germination and protocorm establishment

Experiments using several orchid species have shown that protocorm growth occurs following the formation of the first pelotons. Radioactive tracer studies indicate that the transfer of carbon compounds and phosphorus into growing protocorms occurs via fungal hyphae. Trehalose, a fungal sugar, is translocated to protocorms where it is metabolized into other carbohydrates, including sucrose. Stimulation of protocorm growth during fungal infection appears primarily due to the translocated sugars (see Smith and Read 1997).

Figure 226. Young protocorms (arrowheads) of *Spiranthes sinensis* grown on medium containing ground oats as a carbon source and inoculated with the fungus *Ceratobasidium cornigerum*. Photo courtesy of Dr. Yukari Kuga-Uetake.

Figure 227. Culture plates of two isolates of the fungus, *Ceratobasidium cornigerum*. Photo courtesy of Dr. Yukari Kuga-Uetake.

Figures 228, 229. Cultures of *Epulorhiza* spp. isolated from roots of *Cypripedium reginae*.

Figure 230. Hyphae (arrowheads) and monilioid cells (double arrowheads) of *Epulorhiza* cf. *inquilina* isolated from a root of *Paphiopedilum charlesworthii*.

Figure 231. Runner hypha (arrow) and monilioid cells (arrowheads) of *Epulorhiza repens* isolated from a root of *Paphiopedilum hirsutissimum*.

Photos 228–231 courtesy of Dr. Carla Zelmer.

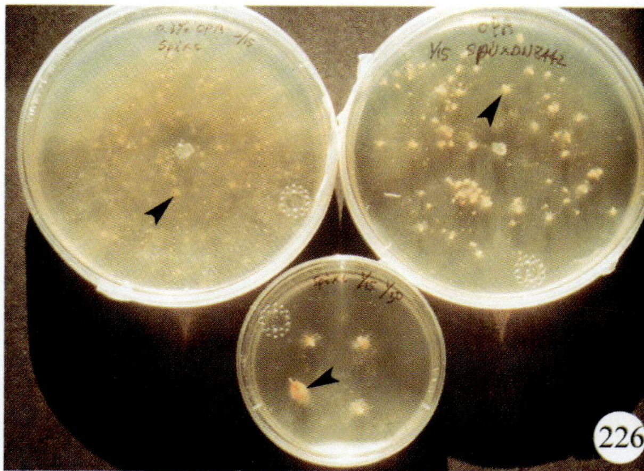
226

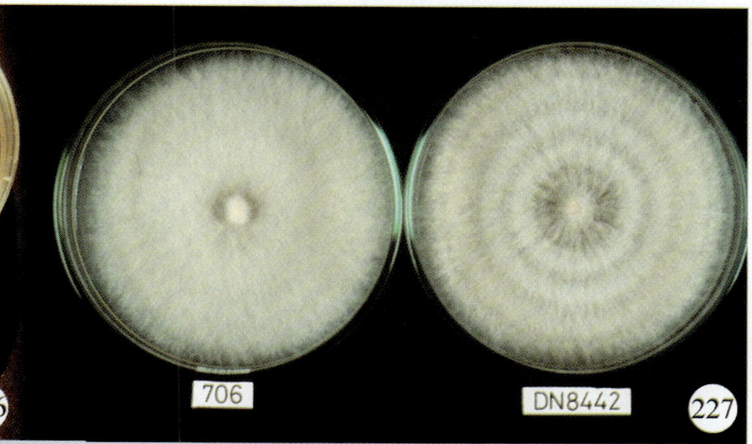
706
DN8442
227

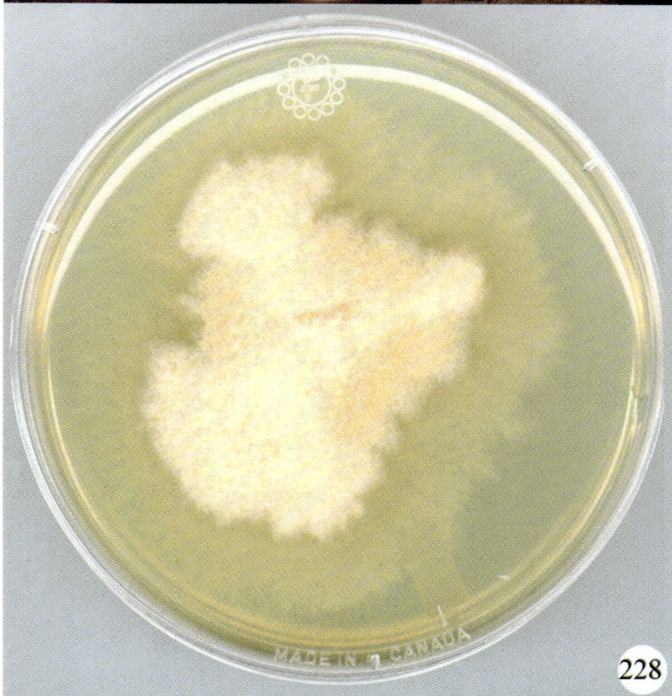
MADE IN CANADA
228

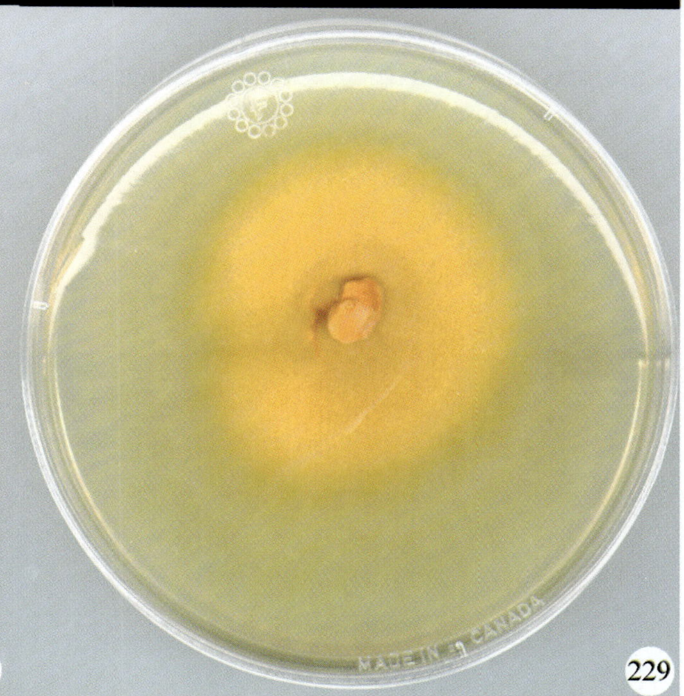
MADE IN CANADA
229

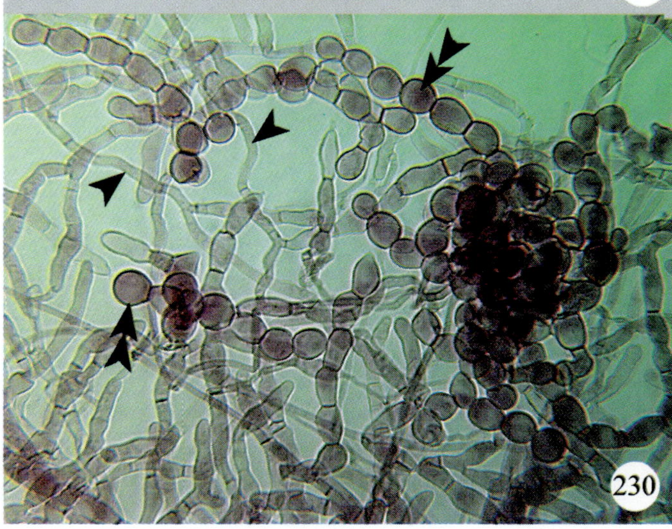
230

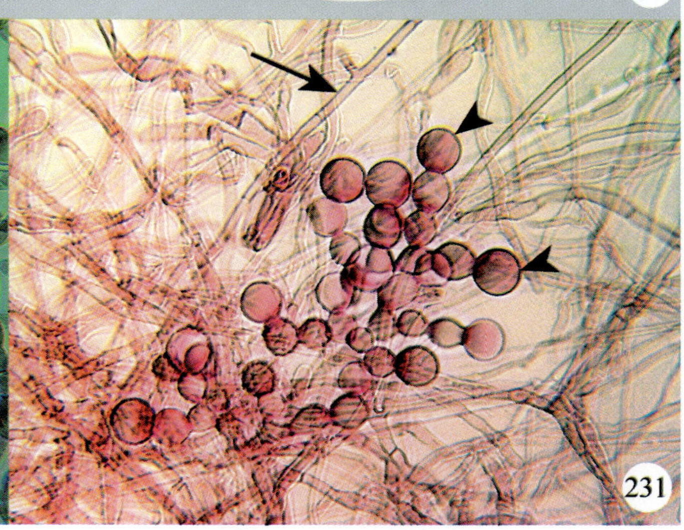
231

Figure 232. 'Dust' seeds of *Spiranthes sinensis*. The probe (arrowhead) is the tip of a fine sewing needle. Photo courtesy of Dr. Yukari Kuga-Uetake.

Figure 233. Scanning electron micrograph of a seed of *Listera australis* showing the pattern of testa cells.

Figure 234. Seed of *Platanthera hyperborea* showing the undifferentiated embryo (*) and epidermal hairs (arrowheads).

Figure 235. Transverse section through the embryo of a *Platanthera hyperborea* seed showing deposits of protein (arrowheads). The outline of the seed coat (testa) is indicated by double arrowheads.

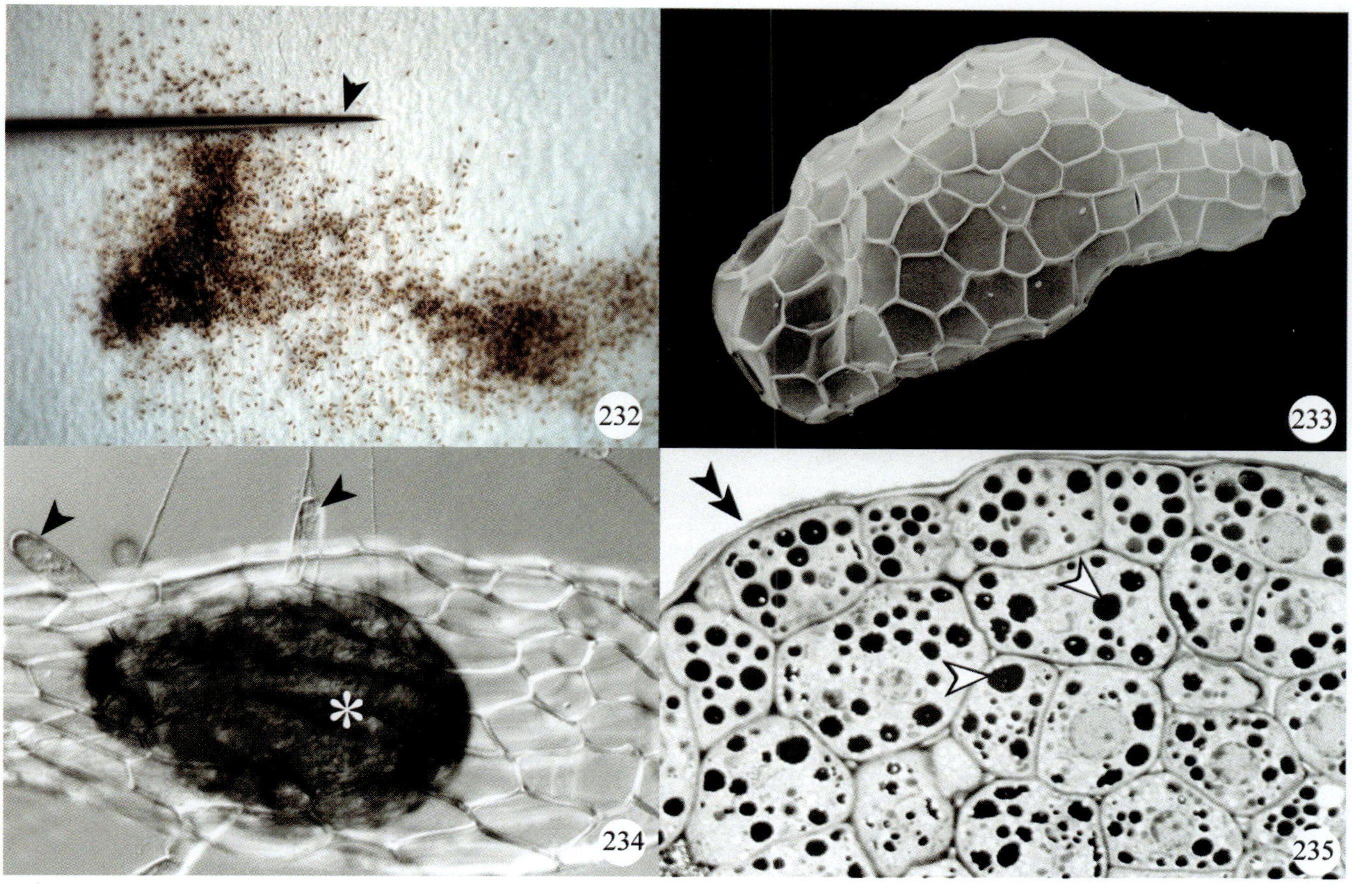
232
233
234
235

Figures 236–239. Diagrammatic representation of the sequence of events from seed to protocorm to seedling for orchid species.

Figure 236. Undifferentiated embryo (*) within an orchid seed.

Figure 237. Colonization of embryo by fungal hyphae (arrowhead), the formation of coils (pelotons, arrows), and the formation of a shoot apical meristem region (*) in a developing protocorm. The testa is not shown in this diagram.

Figure 238. Older stage in protocorm development showing formation of epidermal hairs (arrowheads), and hyphae (double arrowhead) that grow into the substrate. Pelotons (arrow) are evident.

Figure 239. Young seedling with leaves (arrows) and the first adventitious root (R).

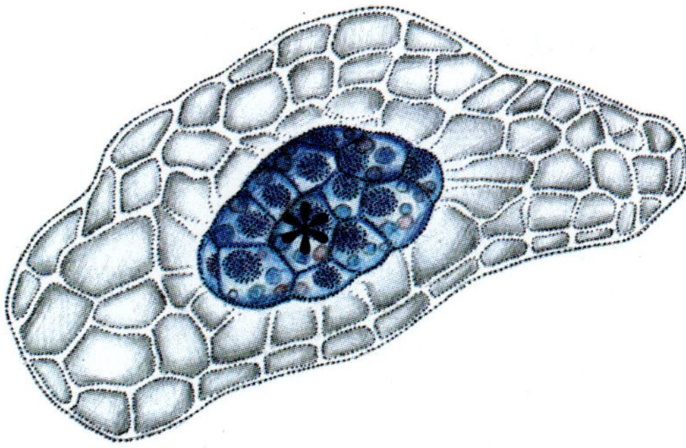

236

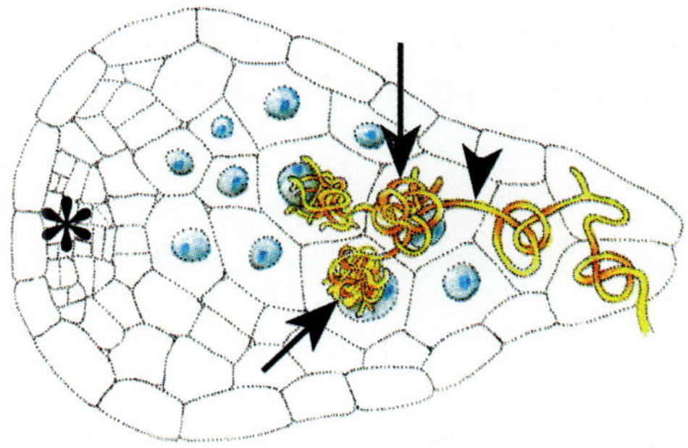

237

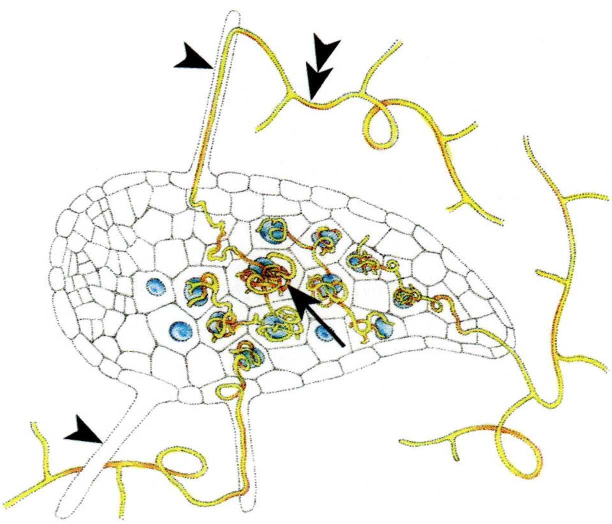

238

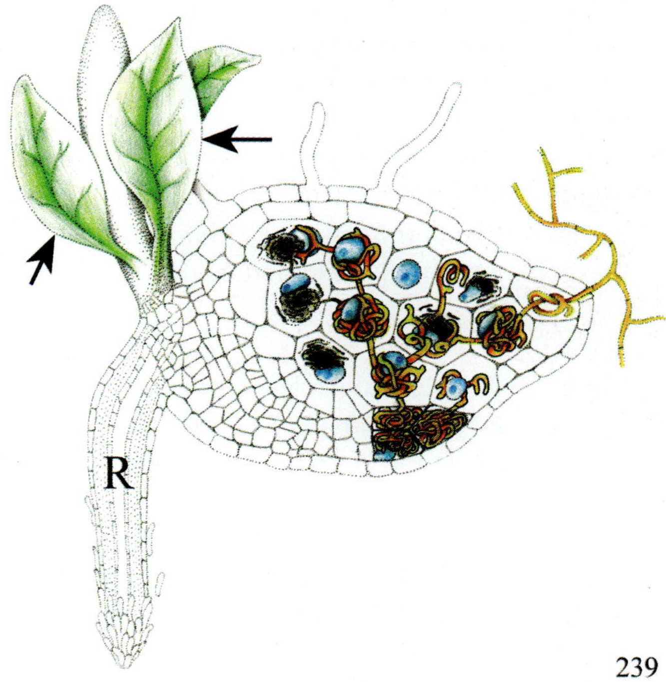

239

Figure 240. Embryo of *Spiranthes sinensis* with a fungal hypha (arrowhead) entering the suspensor end. Dense pelotons (double arrowheads) have formed in parenchyma cells.
Photo courtesy of Dr. Yukari Kuga-Uetake.

Figure 241. Scanning electron micrograph of a protocorm of *Orchis morio* with pelotons (arrowheads).

Figure 242. Transverse section through a protocorm of *Platanthera hyperborea* showing that host cell nuclei (arrowheads) are surrounded by pelotons. Section has been stained with acriflavin-HCl and viewed with blue light.

Figure 243. TEM showing the interface between a peloton hypha (h) of *Ceratobasidium cereale* within a protocorm cell (c) of *Goodyera repens*. The hypha is separated from the host cell cytoplasm by a perifungal membrane (arrowhead) and interfacial matrix material (double arrowhead).

Figure 244. Enlarged nuclei (arrowheads) in protocorm cells of *Spiranthes sinensis* induced by the fungus, *Ceratobasidium cornigerum*. Only cells with pelotons (double arrowheads) show nuclear enlargement. Paraffin-embedded section stained with Schiff's reagent for DNA.

Figure 245. TEM of *Ceratobasidium cereale* hypha penetrating cell walls between adjacent cells in a protocorm of *Goodyera repens*.

Figure 246. Longitudinal section through a protocorm of *Goodyera repens* colonized by *Ceratobasidium cereale*. Several protocorm cells have degenerating pelotons (arrowheads).

Figure 247. Protocorms of *Goodyera repens* grown on medium with cellulose as a carbon source and colonized by *Ceratobasidium cereale*. Several epidermal hairs (arrowheads) have formed.

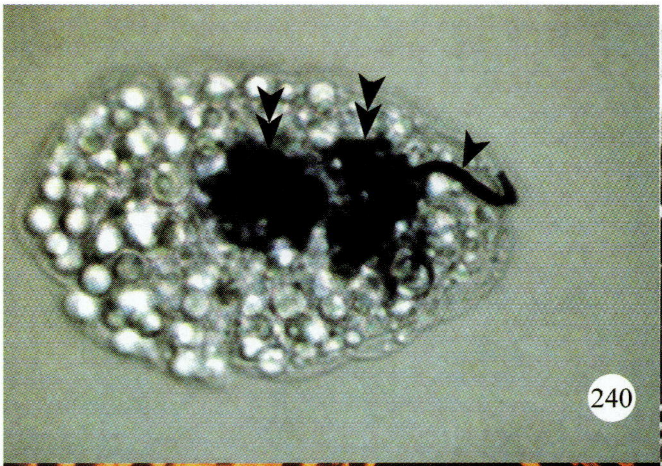
240

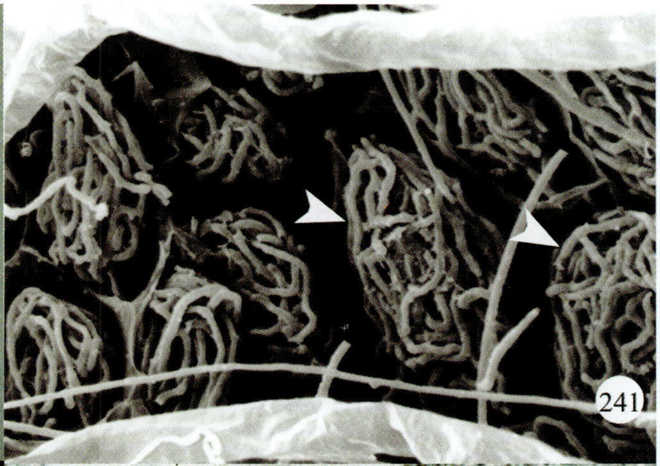
241

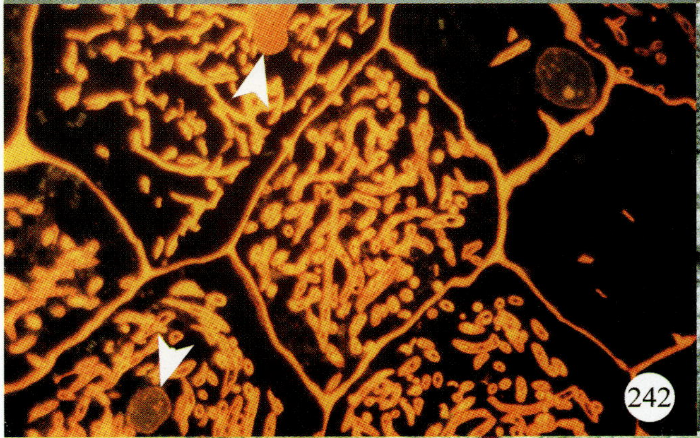
242

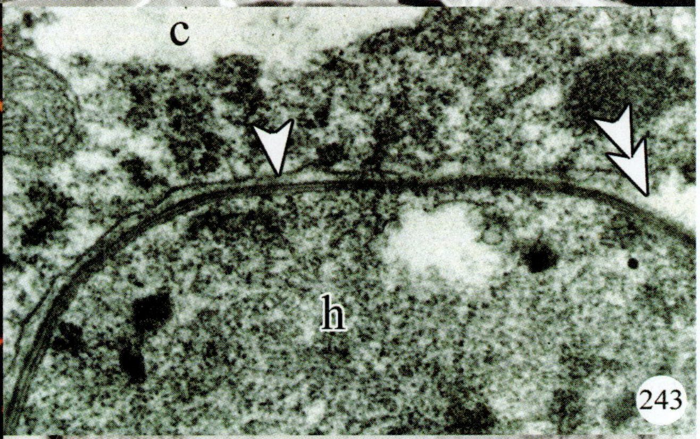
c
h
243

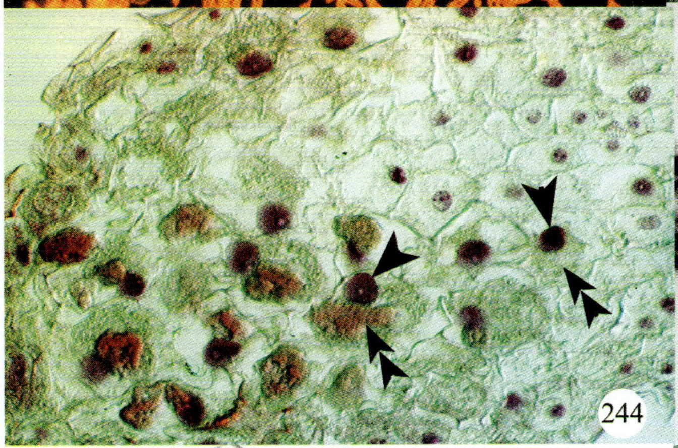
244

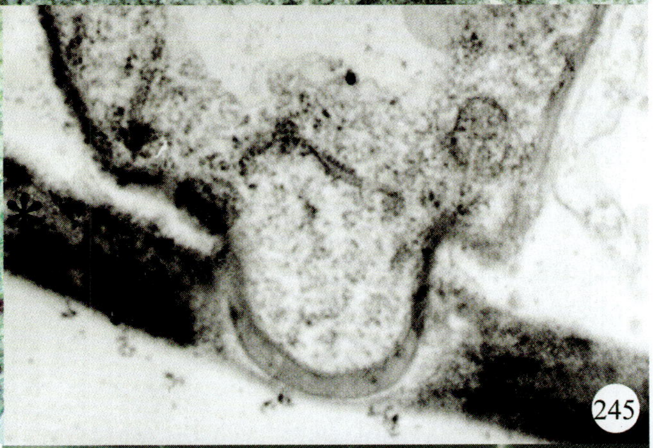
245

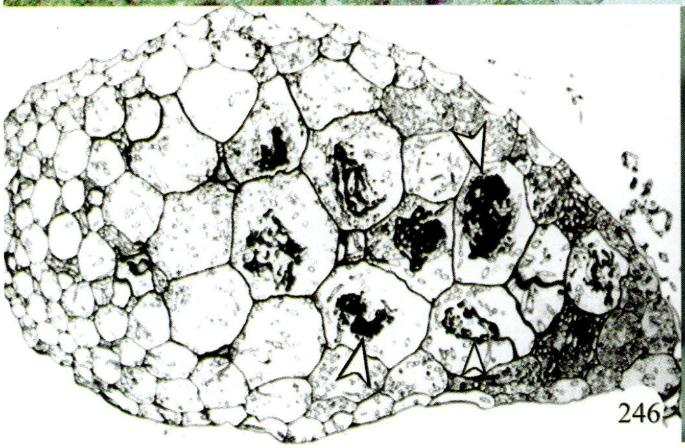
246

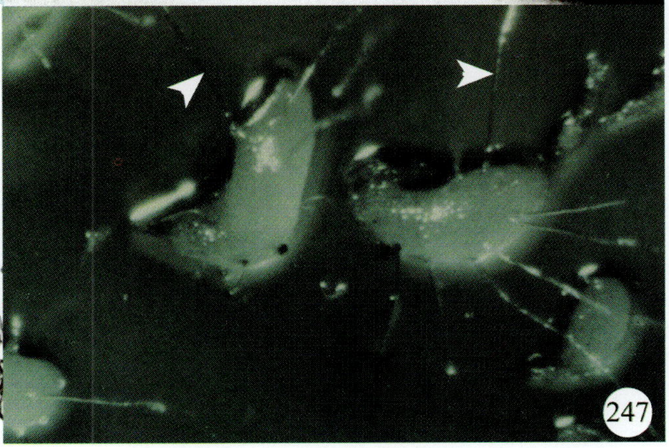
247

Figure 248. Collapsed pelotons of *Ceratorhiza goodyera-repentis*, surrounded by callose (arrowheads) within protocorm cells of *Platanthera hyperborea*.

Figure 249. Collapsed hyphae of *Ceratorhiza goodyera-repentis* in a protocorm cell of *Platanthera hyperborea*. Tissue was labelled with cellobiohydrolase – gold (arrowheads) to reveal cellulose in the interfacial matrix material.

Figure 250. Intact (arrowheads) and degenerated pelotons (arrows) in a root of *Cypripedium*.

Figure 251. Colonization of *Cypripedium arietinum* root. Intact pelotons (arrowheads) and degenerated pelotons (arrows) occur within the same root.

Figures 252, 253. Sections of *Phragmipedium boisserianum* aerial root.

Figure 252. A tilosome (arrowhead) adjacent to a short cell (*) in the exodermis (x). The short cell contains fungal hyphae (arrow). V = Velamen.

Figure 253. Velamen (*) with bacteria and other organisms.

Figure 254. A short cell (arrowheads) in the exodermis of a *Paphiopedilum lawrencianum* aerial root with fungal hyphae. Fungi enter the cortex from the velamen (*) through short cells. Photo taken with crossed polars.

Figure 255. Algae (arrow) and cyanobacteria (arrowheads) on the surface and within the velamen of an aerial root of *Phragmipedium caricinum*.

Figure 256. Seedling of a *Phalaenopsis* species showing fleshy, unbranched roots.

Photos 250–255 courtesy of Dr. Carla Zelmer.

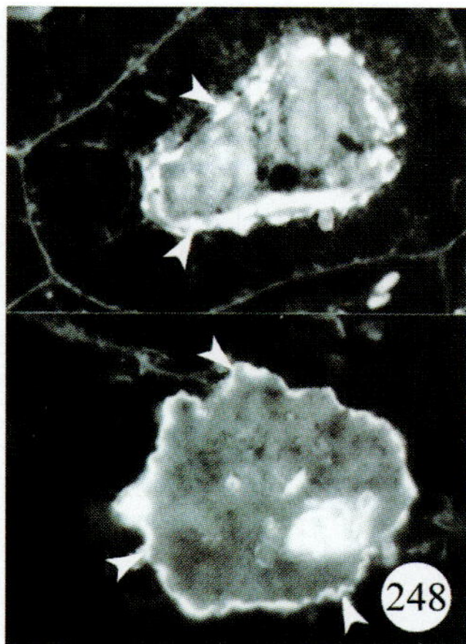
248

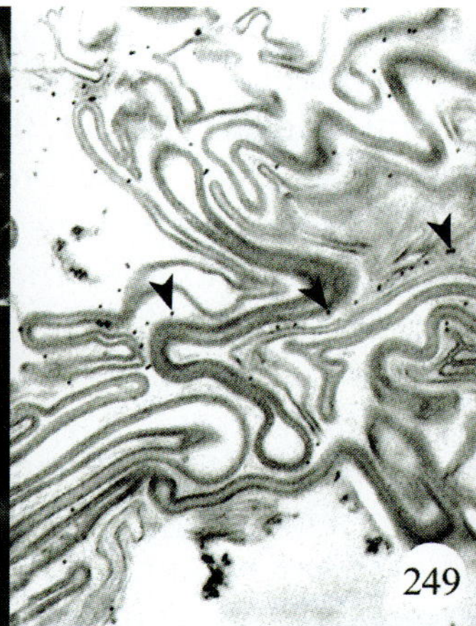
249

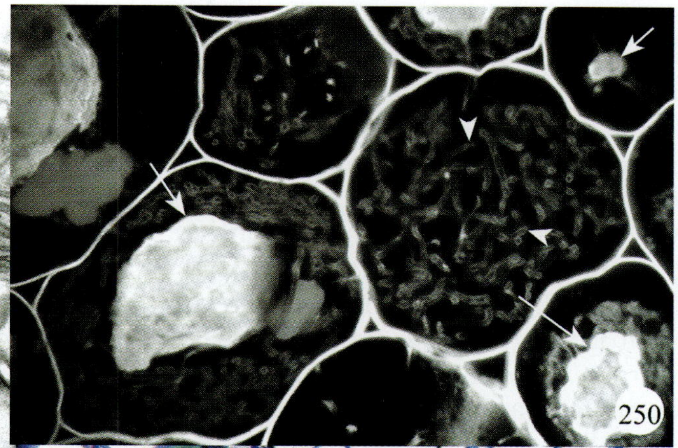
250

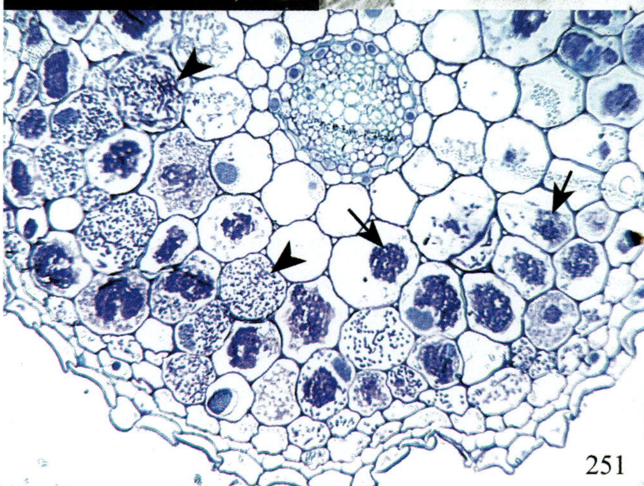
251

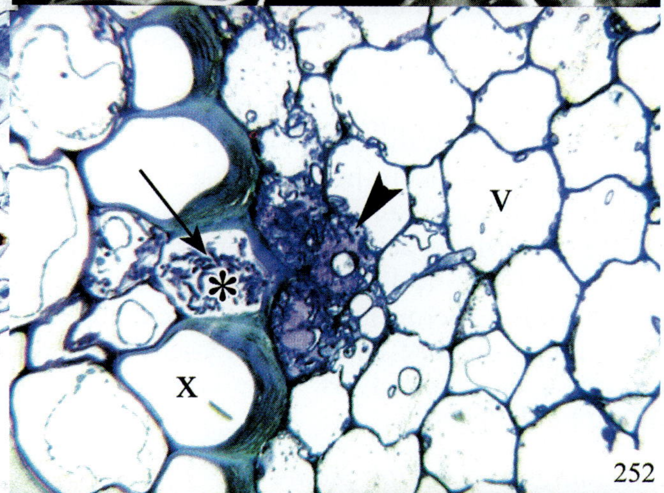
V
*
X
252

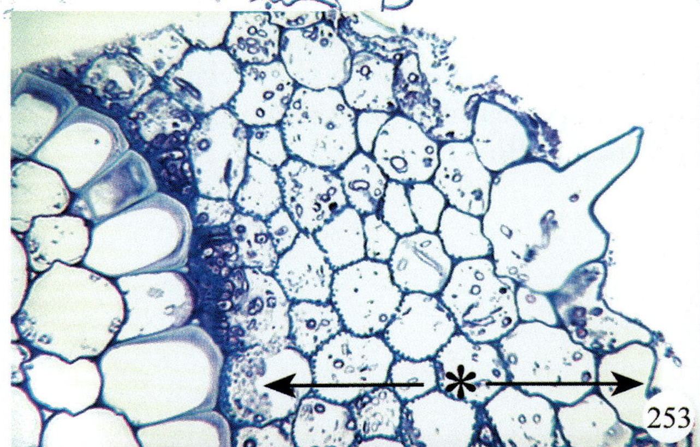
*
253

*
254

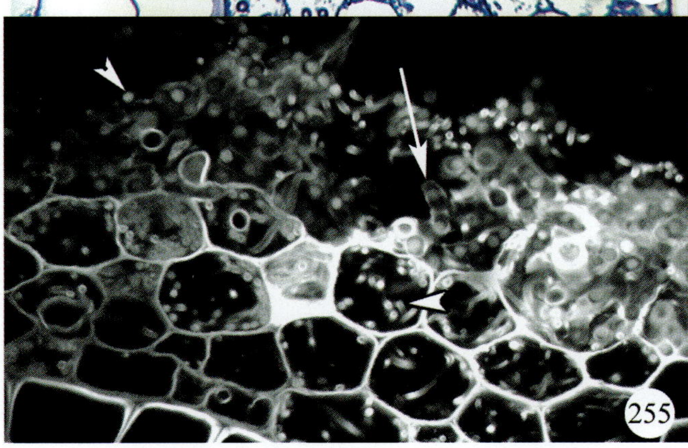
255

256

After exposing fungal mycelium connected to protocorms to a source of radioactive phosphorus, radioactivity was found in protocorm tissues. Although some circumstantial evidence exists regarding the importance of mycorrhizal fungi in nitrogen acquisition by orchids, there is no direct evidence that fungal hyphae transport nitrogen compounds from the soil to the developing protocorms. It is known that the form and concentration of nitrogen provided to germinating seeds and protocorms *in vitro* can have a profound effect on the symbiotic relationship. In some instances, particularly at high levels of nitrogen, the fungus can proliferate and parasitize protocorms resulting in protocorm death (Beyrle et al. 1995).

The site at which sugars and mineral nutrients are transferred from the fungus to the orchid cells is not known. It is likely, however, that since a perifungal membrane surrounds peloton hyphae, transfer of nutrients across this membrane occurs before fungal cell death and subsequent digestion by host cells. Alternately, nutrients accumulated within fungal hyphae could be released into the host cell upon hyphal lysis; these could then be utilized by the plant.

2. Seedlings and mature plants

Although numerous reports document the occurrence of typical orchid mycorrhizal structures in the roots of terrestrial and epiphytic orchids (Rasmussen 1995; Smith and Read 1997), few have considered seedling and mature chlorophyllous plants in terms of the functioning of the mycorrhizal fungi. In fact, there has been some debate as to whether these orchids require mycorrhizal fungi. The morphology of the root systems of chlorophyllous terrestrial orchids, however, is suggestive of the dependency of these plants on fungi for nutrient acquisition. Roots are generally fleshy, unbranched, and few in number (Figure 256), all characteristics of plants with a high dependency on mycorrhizal fungi (Baylis 1975). In addition to questions pertaining to the functional aspects of these mycorrhizas, aspects related to the extent of the extraradical mycelium and its interaction with biotic and abiotic components of the soil need to be explored.

Experimental work with radioactive tracers has shown that *Goodyera repens* seedlings were more dependent than older plants on sugars supplied by the mycorrhizal fungus (Alexander and Hadley 1985). These authors also showed that sugars were not transferred from the plant to the fungus. In the same orchid species, ^{32}P was transported via fungal mycelium to mature plants, and at low concentrations of P, there was enhanced growth of these plants compared to plants whose associated fungal mycelium had been interrupted by a fungicide treatment (Alexander et al. 1984).

Experimental work with myco-heterotrophic orchid species in terms of nutrient acquisition is profiled in **Box 15**.

Box 15: Orchids as 'cheaters'

The relationship between achlorophyllous orchid species and mycorrhizal fungi is complex. Leake (1994) summarized the plant groups that are dependent on fungi for the acquisition of carbon and suggested the term myco-heterotrophic to describe such plants. All orchid species are myco-heterotrophic during seed germination and early seedling establishment and several nonphotosynthetic orchid species are myco-heterotrophic throughout their life cycle. Taylor and Bruns (1999) refer to these latter orchid species as 'cheating' orchids. It has been known for some time that fungi isolated from fully myco-heterotrophic orchid species could form ectomycorrhizas with tree and shrub species (Warcup 1985; Zelmer and Currah 1995); however, more recently the highly specific nature of the association of some of these orchid species with fungal symbionts has been demonstrated using molecular approaches. In the western United States, the orchid species *Cephalanthera austinae* associates only with ectomycorrhizal fungi belonging to the Thelephoraceae clade whereas *Corallorhiza maculata* associates with members of the Russulaceae clade (Taylor and Bruns 1997). These and other myco-heterotrophic orchid species appear to no longer associate with '*Rhizoctonia*' fungi that are able to provide sugars from the breakdown of complex organic compounds in the soil, but rather with ectomycorrhizal fungi that link them to tree host species. Further studies (Taylor and Bruns 1999) showed that, although two species of *Corallorhiza* (*maculata* and *mertensiana*) both associated only with fungal species in the family Russulaceae, they did not share the same fungal species even in populations where they co-occurred. The fungal symbionts in any one specimen did not change with season. There was, however, geographical variation of the fungal symbionts with both orchid species.

In one study, seeds of *Corallorhiza trifida* were enclosed in nylon mesh bags and placed in soil either below stands of *Salix repens* or *Betula–Alnus*; germinating seeds associated only with fungi in the *Thelephora–Tomentella* complex of the Thelephoraceae (McKendrick et al. 2000*a*). In a further study (McKendrick et al. 2000*b*), *C. trifida* seedlings formed hyphal links with roots of *Betula pendula* and *Salix repens* but not with *Pinus sylvestris*. Using $^{14}CO_2$ carbon transfer occurred from *B. pedula* and *S. repens* to *C. trifida,* demonstrating that this orchid species is a 'cheater' by sharing fungal symbionts with tree species.

Figures a, b. Seedlings of the myco-heterotrophic orchid, *Corallorhiza trifida*. **Figure a.** Small seedlings (arrowheads) growing adjacent to roots (arrow) of a *Betula pendula* seedling (B). **Figure b.** The orchid seedling (*) is connected to ectomycorrhizal root tips by hyphae (arrowheads). From McKendrick et al. New Phytol. **145:** 539–548 (2000b).

Box 16: Mycorrhizal fungi alter the cytology of orchid cells

Mycorrhizal fungal hyphae penetrate orchid cell walls but are separated from the cell cytoplasm by the elaboration of the plasma membrane which forms a perifungal membrane around the hyphal coil (peloton). The perifungal membrane has properties that differ from the peripheral plasma membrane; one example is the difference in adenylate cyclase activity (Uetake and Ishizaka 1996). In addition, an interfacial matrix separates the perifungal membrane from the hyphal cell wall, thus providing an apoplastic compartment between the symbionts (Peterson et al. 1996).

As the fungus forms a hyphal coil, both components of the cytoskeleton, the cortical array of microtubules (Uetake et al. 1997), and actin filaments (Uetake and Peterson 1997; 1998) disappear and new cytoskeletal elements become closely associated with the fungal coil. In addition, profiles of endoplasmic reticulum and mitochondria become aggregated around the coil (Uetake and Peterson 2000).

The fungal coil eventually undergoes lysis and, at this time, cellulose and pectins are deposited in the interfacial matrix, effectively walling-off the degenerating hyphal coil from the cytoplasm of the host cell (Peterson et al. 1996). Cortical microtubules and actin filaments reappear in these cells. A re-colonization of cells that contain degraded hyphae frequently occurs.

Another early response of orchid cells to the presence of mycorrhizal fungi is the repeated replication of host cell nuclear DNA with resulting nuclear hypertrophy (Williamson and Hadley 1969). It is not known whether this influences the subsequent development of the symbiosis.

Although there are marked changes in orchid cells as they become colonized by mycorrhizal fungi, the molecular signals involved in these changes have not been investigated.

Figures a–c. Protocorm cells of *Spiranthes sinensis*. **Figure a.** Uncolonized cells showing cortical microtubules (arrowhead). **Figure b.** Cell with a mature peloton showing close association between the peloton formed by the fungus, *Ceratobasidium cornigerum,* and microtubules (arrowhead). **Figure c.** Collapsed peloton with associated microtubules. Cortical microtubules (arrowhead) are now present.

Chapter 8. Dark septate fungal endophytes

Grasslands with occasional trees
such as elm (*Ulmus americana*)
Southern Ontario, Canada
Painted desert
Arizona,
United States
Sub-alpine meadows
McGregor Mountains
Canadian Rockies,
British Columbia
Lupinus sp.
Central British Columbia,
Canada
Matteucia sp.

Chapter 8. Dark septate fungal endophytes

A. Introduction

1. Definition

Endophytic fungi colonize various plant organs (stems, leaves, roots) without causing any apparent deleterious effects to the host. Dark septate endophytes (DSEs) are conidial or sterile ascomycetous fungi that have pigmented (dematiaceous), septate hyphae, and colonize roots of a wide range of plant species (Jumpponen and Trappe 1998). Many of these fungi were originally placed in a heterogeneous group of largely sterile fungi referred to as *Mycelium radicis atrovirens*. Some authors (O'Dell et al. 1993) prefer to call the fungi that colonize roots without causing apparent symptoms, septate endophytes, because some have hyaline rather than dark hyphae.

Although these fungi have been observed for many years in root samples of a variety of plants, only recently have they received much attention in terms of their systematic position and possible functions. We include a discussion of them here because they are widespread and because, under some circumstances, they may also be beneficial to their hosts (Newsham 1999; Jumpponen 2001).

2. Plant species involved

Dark septate fungal endophytes have been recognized in the roots of dicotyledonous and monocotyledonous angiosperms, gymnosperms, ferns, and other seedless vascular plants (Jumpponen and Trappe 1998). These authors provide a list of plant families as well as a web site with over 600 species known to harbour DSEs in their roots. This extensive list indicates the prevalence of this association; in many plant families, only one or a few species have been studied, and often only one collection for a particular species. In most reports, there is no indication of functions played by DSE in particular hosts.

Plant species occurring in cool, nutrionally poor, alpine or subalpine ecosystems, have a high incidence of DSEs (Haselwandter and Read 1980; Read and Haselwandter 1981; Stoyke et al. 1992; Hambleton and Currah 1997). In their study of fungal endophytes in members of the Ericaceae, Hambleton and Currah (1997) found that two species in the ericaceous genus *Vaccinium* growing in sand dunes were always colonized by the dark septate endophyte, *Phialocephala fortinii*. *P. fortinii* is tolerant of a range of soil moisture levels varying from sand dunes to bogs (Addy et al. 2000). In *Atriplex canescens* (fourwing saltbush), a dominant and ecologically important shrub occurring in arid, southwestern rangelands in the United States and southern Canada, roots are consistently colonized by DSEs (Barrow and Aaltonen 2001). It is not uncommon for plant species to have typical mycorrhizal associations in addition to having roots colonized by DSE fungi. For example, many orchid species are symbiotic with mycorrhizal fungi and also associate with DSEs (Currah et al. 1997).

Although uncertain, DSE fungi do not appear to be host specific. For example, an isolate of *Phialocephala fortinii*, originally isolated from roots of *Vaccinium vitis-idaea*, formed typical structures within roots of *Asparagus officinalis* (Yu et al. 2001b). In another study, an unidentified DSE isolated from roots of *Ranunculus adoneus* formed a typical association with roots of *Zea mays* (Schadt et al. 2001).

3. Fungal species involved

Identifying DSEs and other septate fungal root endophytes is problematic because most isolates obtained from roots remain sterile when grown in culture. A few isolates have formed conidia and spores in culture and these, along with other features of the cultured mycelium, can be useful in species identification (Currah and Tsuneda 1993). More recently, molecular techniques have improved our ability to identify some of these root endophytes (Stoyke et al. 1992; Hambleton and Currah 1997; Jumpponen and Trappe 1998; Schadt et al. 2001; Grünig et al. 2002). Although it is certain that there are many unidentified DSEs and other root endophytes, five main species, *Chloridium paucisporum*, *Leptodontidium orchidicola*, *Phialocephala dimorphospora*, *Phialocephala fortinii*, and *Phialophora finlandia* have been identified from a number of plant species (see Jumpponen and Trappe 1998).

The most abundant species identified by molecular methods in roots of subalpine plants growing in Alberta, Canada, was *Phialocephala fortinii* (Stoyke et al. 1992). *P. fortinii* was also the most common root endophyte occurring in non-ectomycorrhizal fine roots of 14 tree and shrub species collected in Switzerland, Finland and Germany (Ahlich and Sieber 1996).

Phialocephala fortinii has been studied extensively in axenic culture (Currah and Tsuneda 1993). Cultures are darkly pigmented (Figure 257) and some of the septate hyphae change from being thin-walled and hyaline to thick-walled and melanized (Figure 258). Hyphae often have pigmented exudates on their surface as well as verrucose walls (Figure 258). Coiled hyphae, hyphal strands (Figures 259, 260), and sclerotium-like structures (Figures 261, 262) often develop in culture. Cultures placed in the dark at 3–5 °C may form conidia, asexual reproductive structures.

B. Colonization of roots

Based on the few studies that have been reported, there appears to be a fairly common pattern of root colonization by dark septate endophytes. Fungal hyphae contact the root surface (Figure 263) and frequently form a network of mycelium over the surface and within the root (Figure 264). Hyphae enter the root either through root hairs, through epidermal cells, through root cap cells if root apical meristems are invaded, or through wounds caused by the emergence of lateral roots. Once inside the root, hyphae proliferate and invade epidermal and cortical cells (Figures 265, 266).

There are some reports in the literature of invasion of vascular tissues as well. Narrow hyaline hyphae develop but are difficult to locate unless either the lipid bodies within them are stained (Figure 267), or roots are stained with a fluorescent dye that can be detected by confocal microscopy (Figures 265, 266). Hyaline hyphae can become melanized and frequently form microsclerotia (compact groups of hyphae) within epidermal and cortical cells (Figure 268). These latter structures, because of their prevalence, can be used as a diagnostic feature of DSEs in field-collected material. Microsclerotia contain polysaccharides (Figure 269), proteins (Figure 270), and polyphosphates (Figure 271) (Yu et al. 2001*b*), strengthening the view that these may be perennating bodies that allow the fungus to survive in older roots in the soil, and that act as inoculum for colonization of new roots. This has not, however, been confirmed.

C. Functions

In spite of the ubiquitous nature of dark septate root endophytes, there is little information as to their functions. A review by Jumpponen (2001) summarizes the experimental data related to the effect of known DSE species on plant growth; it includes a discussion of the evidence for and against DSEs being considered as mycorrhizal fungi. Studies on *Phialocephala fortinii*, show effects on plant growth that range from negative through neutral to positive, depending on host species and possibly the conditions under which the plants are grown. There is no direct experimental evidence demonstrating that DSEs are involved in nutrient acquisition although indirect evidence supports this view. An increase in shoot phosphorus was observed in two *Carex* species (Haselwandter and Read 1982) as well as in *Pinus contorta* (Jumpponen et al. 1998) when these plants were inoculated with *P. fortinii*. In addition, the annual grass species *Vulpia ciliata* ssp. *ambigua* showed increased root, shoot and total seedling phosphorus levels when plants were colonized by the dark septate fungus, *Phialophora graminicola* (Newsham 1999). Positive growth effects also occurred in the colonized grass plants.

There is some evidence that *P. fortinii*, as well as other root endophytes, may be useful as biocontrol organisms. In one study, axenically-grown seedlings of eggplant (*Solanum melongena*) pre-inoculated with *P. fortinii* were protected from the pathogenic wilt fungus, *Verticillium dahliae* (Narisawa et al. 2002).

Functionally, it has been demonstrated that DSE fungi are able to utilize cellulose, protein, organic nitrogen, starch, RNA, and xylans (see Caldwell et al. 2000). It is possible that these fungi may provide plants with access to more recalcitrant pools of carbon, nitrogen and phosphorus in litter and other detritus. This source of nutrition may be important in soils in which phosphorus and nitrogen are bound to organic compounds and are, therefore, not immediately available. Theoretically, seedling establishment could benefit from associations with DSEs through enhanced nutrient availability. Recently, Bartholdy et al. (2001) demonstrated that isolates of *P. fortinii* secrete siderophores into culture medium, suggesting that this fungus may be able to compete with other soil organisms for iron.

Further experimental studies are needed before definite conclusions can be reached concerning any benefits that plants may gain by being associated with DSEs. Jumpponen (2001) suggests that long-term benefits might result from these associations, but this has not yet been demonstrated.

Figures 257–262. Features of *Phialocephala fortinii* cultures.

Figure 257. The dematiaceous nature of the mycelium grown on Potato Dextrose Agar.

Figure 258. Hyaline (large arrows) and melanized (double arrowhead) hyphae. Septa (arrowheads) and exudates (small arrows) are evident.

Figures 259, 260. Hyphal strands (arrowheads) formed in cultured mycelium.

Figures 261, 262. Sclerotium-like structures (arrowheads) formed in cultured mycelium.

Figure 263. Early stage in colonization of an *Asparagus officinalis* root by hyphae (arrowheads) of *Phialocephala fortinii* viewed by scanning electron microscopy.

Figure 264. Surface hyphae, internal hyphae and microsclerotia (arrowheads) of a dark septate endophyte from field-collected roots of a grass species. Photo courtesy of Leon Commandeur.

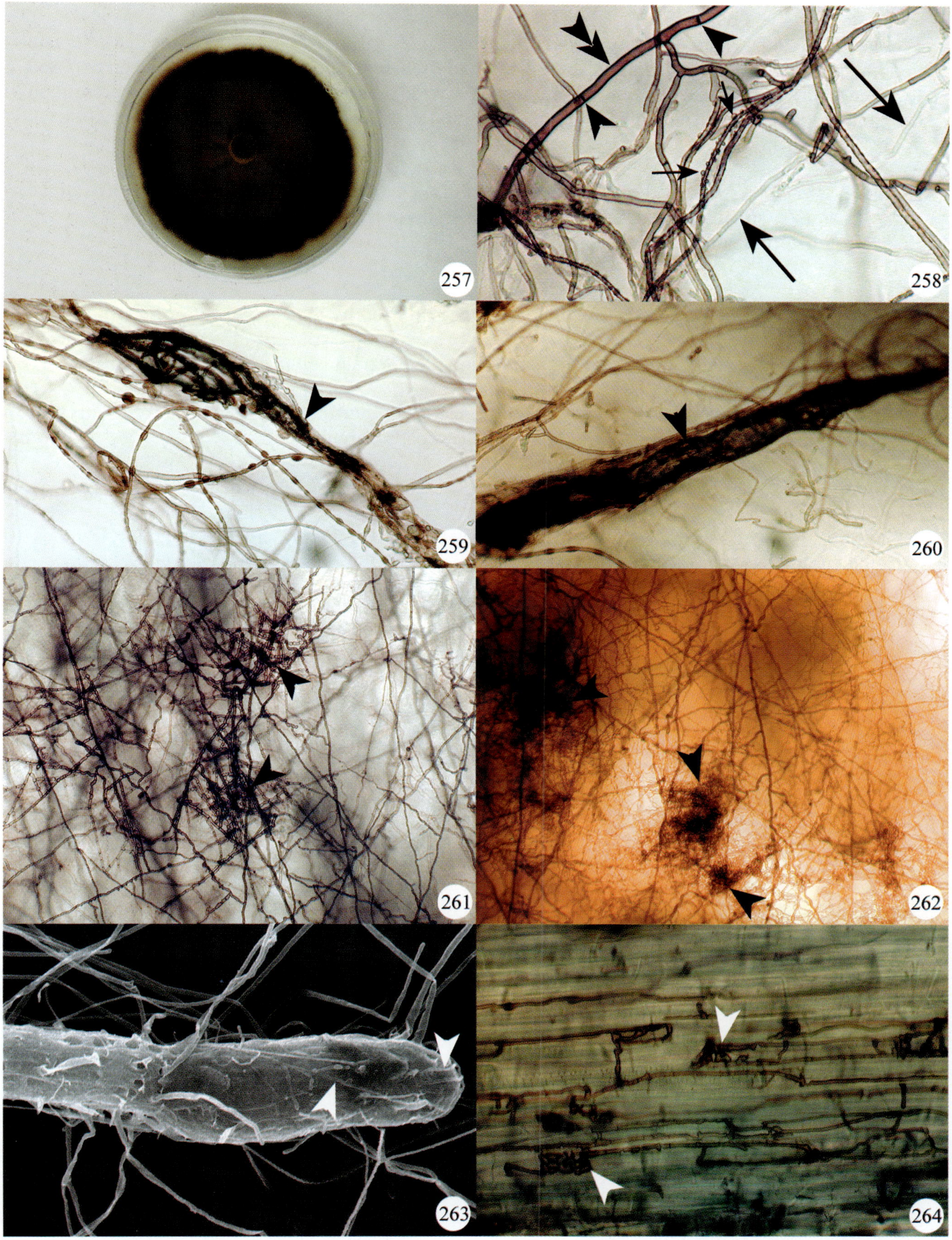
257
258
259
260
261
262
263
264

Figures 265–266. Cleared roots of *Populus tremuloides* (aspen) colonized by the dark septate endophyte, *Phialocephala fortinii,* stained with acid fuchsin (a stain that fluoresces when exposed to green light) and viewed by confocal microscopy. Wide diameter hyphae (arrows) are present on the root surface and narrow hyphae with many septa (arrowheads) are present within the cortex.

Figure 267. Dark septate endophyte hyphae within a cleared root of *Atriplex canescens* stained with Sudan IV. Numerous lipid bodies (arrowheads) are present in the hyphae.
Photo from Barrow and Aaltonen. Mycorrhiza **11:** 199–205 (2001).

Figures 268–271. Microsclerotia of *Phialocephala fortinii* in roots of *Asparagus officinalis*.

Figure 268. Melanized microsclerotium within an epidermal cell of a cleared root.

Figure 269. Section of a microsclerotium stained with acriflavin-HCl and viewed with blue light. Polysaccharides, probably glycogen (arrowheads), are present.

Figure 270. Section stained with amido black showing proteins (arrowheads).

Figure 271. Section stained with DAPI and viewed with ultra violet light. Numerous polyphosphate granules (arrowheads) are present.

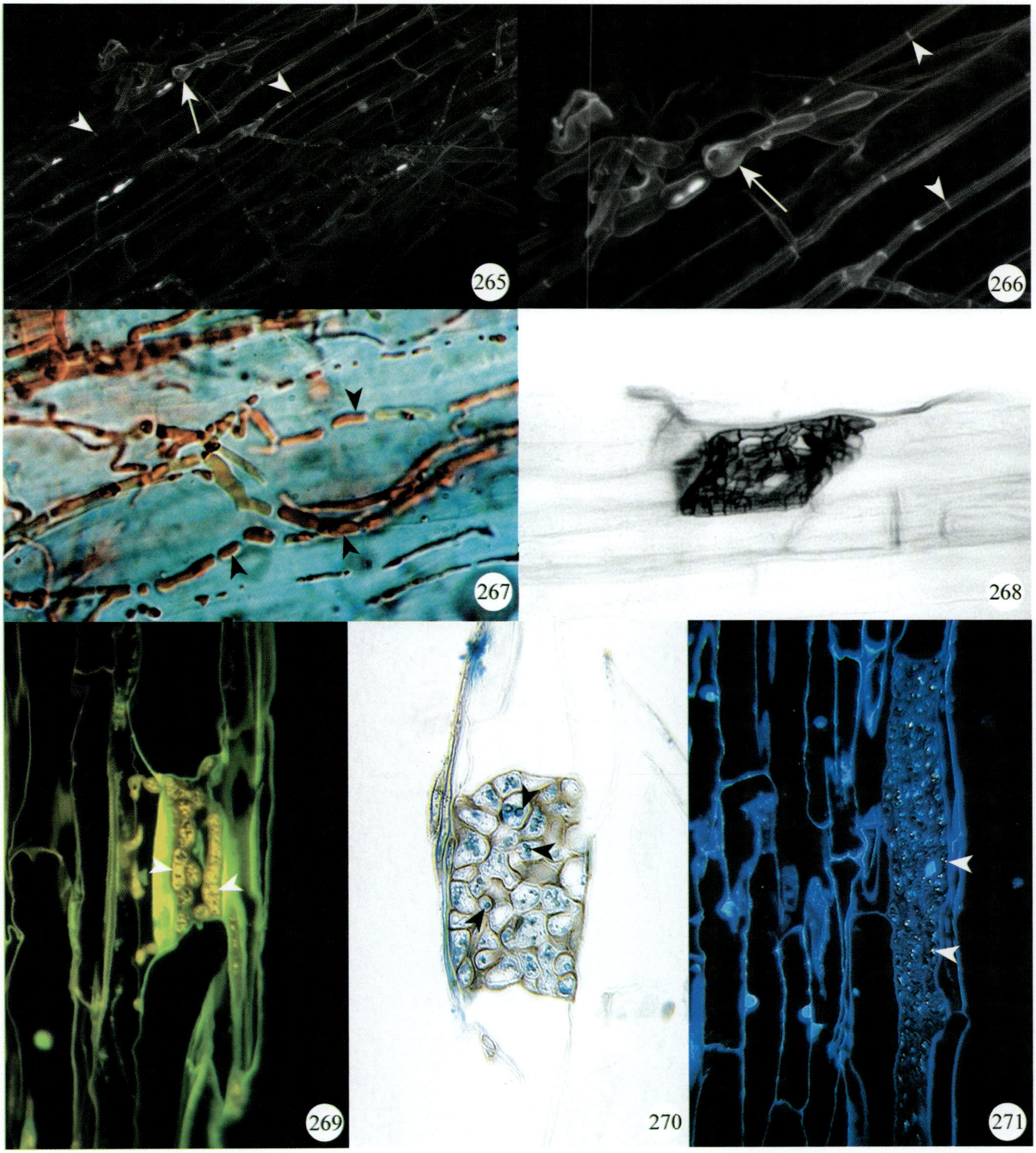
265
266
267
268
269
270
271

References

Acsai, J., and Largent, D.L. 1983*a*. Fungi associated with *Arbutus menziesii*, *Arctostaphylos uva-ursi* in central and northern California. Mycologia, **75:** 544–547.

Acsai, J., and Largent, D.L. 1983*b*. Mycorrhizae of *Arbutus menziesii* Pursh and *Arctostaphylos manzanita* Parry in northern California. Mycotaxon, **16:** 519–536.

Addy, H.D., Hambleton, S., and Currah, R.S. 2000. Distribution and molecular characterization of the root endophyte *Phialocephala fortinii* along an environmental gradient in the boreal forest of Alberta. Mycol. Res. **104:** 1213–1221.

Agerer, R. 1992. Ectomycorrhizal rhizomorphs: organs of contact. *In* Mycorrhizas in ecosystems. *Edited by* D.J. Read, D.H. Lewis, A.H. Fitter and I.J. Alexander. CAB International, Wallingford, UK. pp. 84–90.

Agerer, R. 1987–2002. Colour Atlas of Ectomycorrhizae. Einhorn-Verlag, Schwäbisch Gmünd, Germany.

Ahlich, K., and Sieber, T.N. 1996. The profusion of dark septate endophytic fungi in non-ectomycorrhizal fine roots of forest trees and shrubs. New Phytol. **132:** 259–270.

Alexander, C., Alexander, I.J., and Hadley, G. 1984. Phosphate uptake in relation to mycorrhizal infection. New Phytol. **97:** 401–411.

Alexander, C., and Hadley, G. 1985. Carbon movement between host and mycorrhizal endophyte during the development of the orchid *Goodyera repens* Br. New Phytol. **101:** 657–665.

Allaway, W.G., and Ashford, A.E. 2001. Motile tubular vacuoles in extramatrical mycelium and sheath hyphae of ectomycorrhizal systems. Protoplasma, **215:** 218–225.

Allaway, W.G., Carpenter, J.L., and Ashford, A.E. 1985. Amplification of inter-symbiont surface by root epidermal transfer cells in the *Pisonia* mycorrhiza. Protoplasma, **128:** 227–231.

Armstrong, L., and Peterson, R.L. 2002. The interface between the arbuscular mycorrhizal fungus *Glomus intraradices* and root cells of *Panax quinquefolius* : a *Paris*-type mycorrhizal association. Mycologia, **94:** 587–595.

Arocena, J.M., Glowa K.R., Massicotte, H.B., and Lavkulich, L. 1999. Chemical and mineral composition of ectomycorrhizosphere soils of subalpine fir (*Abies lasiocarpa* (Hook.) Nutt.) in the Ae horizon of a luvisol. Can. J. Soil Sci. **79**: 25–35.

Ashford, A.E., and Allaway, W.G. 1982. A sheathing mycorrhiza on *Pisonia grandis* R. Br. (Nyctaginaceae) with development of transfer cells rather than a Hartig net. New Phytol. **90:** 511–519.

Ashford, A.E., and Allaway, W.G. 1985. Transfer cells and Hartig net in the root epidermis of the sheathing mycorrhiza of *Pisonia grandis* R.Br. from Seychelles. New Phytol. **100:** 595–612.

Ashford, A.E., Allaway, W.G., and Reed, M.L. 1996. A possible role for the thick-walled epidermal cells in the mycorrhizal hair roots of *Lysinema ciliatum* R.Br., and other Epacridaceae. Ann. Bot. **77:** 375–381.

Ashford, A.E., Cole, L., and Hyde, G.J. 2001. Motile tubular vacuole systems. *In* The Mycota VIII Biology of the Fungal Cell. *Edited by* R.J. Howard and N.A.R Gow. Springer-Verlag, Berlin. pp. 243–265.

Ashford, A.E., Ling Lee, M., and Chilvers, G.A. 1975. Polyphosphate in eucalypt mycorrhizas: a cytochemical demonstration. New Phytol. **74:** 447–453.

Ashford, A.E., Peterson, R.L., Dwarte, D., and Chilvers, G.A. 1986. Polyphosphate granules in eucalypt mycorrhizas: determination by energy dispersive x-ray analysis. Can. J. Bot. **64:** 677–687.

Ashford, A.E., Vesk, P.A., Orlovich, D.A., Markovina, A.-L., and Allaway, W.G. 1999. Dispersed polyphosphate in fungal vacuoles in *Eucalyptus pilularis/Pisolithus tinctorius* ectomycorrhizas. Fungal Genet. Biol. **28:** 21–33.

Bago, B., Azcón-Aguilar C., Goulet, A., and Piché, Y. 1998*a*. Branched absorbing structures (BAS): a feature of the extraradical mycelium of symbiotic arbuscular mycorrhizal fungi. New Phytol. **139:** 375–388.

Bago, B., Azcón-Aguilar, C., and Piché, Y. 1998*b*. Architecture and developmental dynamics of the external mycelium of the arbuscular mycorrhizal fungus *Glomus intraradices* grown under monoxenic conditions. Mycologia, **90:** 52–62.

Bago, B., Vierheilig, H., Piché, Y., and Azcón-Aguilar, C. 1996. Nitrate depletion and pH changes induced by the extraradical mycelium of the arbuscular mycorrhizal fungus *Glomus intraradices* grown in monoxenic culture. New Phytol. **133:** 273–280.

Bais, H.P., Loyola-Vargas, V.M., Flores, H.E., and Vivanco, J.M. 2001. Root-specific metabolism: the biology and biochemistry of underground organs. *In Vitro* Cell. Dev. Biol.- Plant. **37:** 730–741.

Bajwa, R., Abuarghub, S., and Read, D.J. 1985. The biology of mycorrhiza in the Ericaceae X. The utilization of proteins and the production of proteolytic enzymes by the mycorrhizal endophyte and by mycorrhizal plants. New Phytol. **101:** 469–486.

Bajwa, R., and Read, D.J. 1985. The biology of mycorrhiza in the Ericaceae. IX. Peptides as nitrogen sources for the ericoid endophyte and for mycorrhizal and non-mycorrhizal plants. New Phytol. **101:** 459–468.

Bajwa, R., and Read, D.J. 1986. Utilization of mineral and amino N sources by the ericoid mycorrhizal endophyte *Hymenoscyphus ericae* and by mycorrhizal and non-mycorrhizal seedlings of *Vaccinium*. Trans. Br. Mycol. Soc. **87:** 269–277.

Balaji, B., Bâ, A.M., LaRue, T.A., Tepfer, D., and Piché, Y. 1994. *Pisum sativum* mutants insensitive to nodulation are also insensitive to invasion *in vitro* by the mycorrhizal fungus, *Gigaspora margarita*. Plant Science, **102:** 195–203.

Balaji, B., Poulin, M.J., Vierheilig, H., and Piché, Y. 1995. Responses of an arbuscular mycorrhizal fungus, *Gigaspora margarita*, to exudates and volatiles from the Ri T-DNA-transformed roots of nonmycorrhizal and mycorrhizal mutants of *Pisum sativum* L. Sparkle. Exp. Mycol. **19:** 275–283.

Balestrini, R., Romera, C., Puigdomenech, P., and Bonfante, P. 1994. Location of a cell wall hydroxyproline-rich glycoprotein, cellulose and β-1, 3 glucans in apical and differentiated regions of maize mycorrhizal roots. Planta, **195:** 201-209.

Barker, S.J., Stummer, B., Gao, L., Dispain, I., O'Connor, P.J., and Smith, S.E. 1998. A mutant in *Lycopersicon esculentum* Mill. with highly reduced VA mycorrhizal colonization: isolation and preliminary characterization. The Plant J. **15:** 791–797.

Barrow, J.R., and Aaltonen, R.E. 2001. Evaluation of the internal colonization of *Atriplex canescens* (Pursh) Nutt. roots by dark septate fungi and the influence of host physiological activity. Mycorrhiza, **11:** 199–205.

Bartholdy, B.A., Berreck, M., and Haselwandter, K. 2001. Hydroxamate siderophore synthesis by *Phialocephala fortinii*, a typical dark septate fungal root endophyte. BioMetals, **14:** 33–42.

Baylis, G.T.S. 1975. The magnolioid mycorrhiza and mycotrophy in root systems derived from it. *In* Endomycorrhizas. *Edited by* F.E. Sanders, B. Mosse, and P.B. Tinker. Academic Press, London, UK. pp. 373–389.

Bécard, G., and Fortin, J.A. 1988. Early events of vesicular–arbuscular mycorrhiza formation on Ri T-DNA transformed roots. New Phytol. **108:** 211–218.

Bécard, G., and Piché, Y. 1989*a*. New aspects on the acquisition of biotrophic status by a vesicular–arbuscular mycorrhizal fungus, *Gigaspora margarita*. New Phytol. **112:** 77–83.

Bécard, G., and Piché, Y. 1989*b*. Fungal growth stimulation by CO_2 and root exudates in vesicular–arbuscular mycorrhizal symbiosis. Appl. Environ. Microbiol. **55:** 2320–2325.

Behrmann, P., and Heyser, W. 1992. Apoplastic transport through the fungal sheath of *Pinus sylvestris/Suillus bovinus* ectomycorrhizas. Botanica Acta, **105:** 427–434.

Bell, T.L., and Pate, J.S. 1996. Nitrogen and phosphorus nutrition in mycorrhizal Epacridaceae of South-west Australia. Ann. Bot. **77:** 389–397.

Bel Rhild, R., Chabot, S., Piché, Y., and Chênevert, R. 1993. Isolation and identification of flavonoids from Ri T-DNA-transformed roots (*Daucus carota*) and their significance in vesicular–arbuscular mycorrhiza. Phytochemistry, **33:** 1369–1371.

Bending, G.D., and Read, D.J. 1996. Nitrogen mobilization from protein-polyphenol complex by ericoid and ectomycorrhizal fungi. Soil Biol. Biochem. **28:** 1603–1612.

Bending, G.D., and Read, D.J. 1997. Lignin and soluble phenolic degradation by ectomycorrhizal and ericoid mycorrhizal fungi. Mycol. Res. **101:** 1348–1354.

Benhamou, N., Fortin, J.A., Hamel, C., St-Arnaud, M., and Shatilla, A. 1994. Resistance responses of mycorrhizal Ri T-DNA-transformed carrot roots to infection by *Fusarium oxysporum* f.sp. *chrysanthemi.* Phytopathology **84:** 958–968.

Berch, S.M., Allen, T.R., and Berbee, M.L. 2002. Molecular detection, community structure and phylogeny of ericoid mycorrhizal fungi. Plant Soil, **244:** 55–66.

Beyrle, H.F., Smith, S.E., Peterson, R.L., and Franco, C.M.M. 1995. Colonization of *Orchis morio* protocorms by a mycorrhizal fungus: effects of nitrogen nutrition and glyphosate in modifying the responses. Can. J. Bot. **73:** 1128–1140.

Bianciotto, V., Bandi, C., Minerdi, D., Sironi, M., Tichy, H.V., and Bonfante, P. 1996. An obligately endosymbiotic fungus itself harbors obligately intracellular bacteria. Appl. Environ. Microbiol. **62:** 3005–3010.

Bianciotto, V., Lumini, E., Bonfante, P., and Vandamne, P. 2003. '*Candidatus Glomeribacter gigasporarum* ' gen. nov., sp. nov., an endosymbiont of arbuscular mycorrhizal fungi. Int. J. Syst. Evol. Microbiol. **53:** 121–124.

Bianciotto, V., Lumini, E., Lanfranco, L., Minerdi, D., Bonfante, P., and Perotto, S. 2000. Detection and identification of bacterial endosymbionts in arbuscular mycorrhizal fungi belonging to the family Gigasporaceae. Appl. Environ. Microbiol. **66:** 4503–4509.

Bidartondo, M.I., and Bruns, T.D. 2001. Extreme specificity in epiparasitic Monotropoideae (Ericaceae): widespread phylogenetic and geographical structure. Mol. Ecol. **10:** 2285–2295.

Bidartondo, M.I., Redecker, D., Hijri, I., Wiemken, A., Bruns, T.D., Domínguez, L., Sérsic, A., Leake, J.R., and Read, D.J. 2002. Epiparasitic plants specialized on arbuscular mycorrhizal fungi. Nature, **419:** 389–392.

Björkman, E. 1960. *Monotropa hypopithys* L. – an epiparasite on tree roots. Physiol. Plant. **13:** 308–327.

Blancaflor, E.B., Zhao, L., and Harrison, M.J. 2001. Microtubule organization in root cells of *Medicago truncatula* during development of an arbuscular mycorrhizal symbiosis with *Glomus versiforme.* Protoplasma, **217:** 154–165.

Boisson-Dernier, A., Chabaud, M., Garcia, F., Bécard, G., Rosenberg, C., and Barker, D.G. 2001. *Agrobacterium rhizogenes*-transformed roots of *Medicago truncatula* for the study of nitrogen-fixing and endomycorrhizal symbiotic associations. MPMI **14:** 695–700.

Bonfante, P., Balestrini, R., and Mendgen, K. 1994. Storage and secretion processes in the spore of *Gigaspora margarita* Becker & Hall as revealed by high-pressure freezing and freeze substitution. New Phytol. **128:** 93–101.

Bonfante, P., Genre, A., Faccio, A., Martini, I., Schauser, L., Stougaard, J., Webb, J., and Parniske, M. 2000. The *Lotus japonicus LjSym4* Gene is required for the successful symbiotic infection of root epidermal cells. MPMI **13:** 1109–1120.

Bonfante, P., and Perotto, S. 1995. Strategies of arbuscular mycorrhizal fungi when infecting host cells. New Phytol. **130:** 3–21.

Bradley, R., Burt, A.J., and Read, D.J. 1982. The biology of mycorrhiza in the Ericaceae VIII. The role of mycorrhizal infection in heavy metal resistance. New Phytol. **91:** 197–209.

Brundrett, M., Bougher, N., Dell, B., Grove, T., and Malajczuk, N. 1996. Working with mycorrhizas in forestry and agriculture. Australian Centre for International Agricultural Research Monograph 32. Pirie Printers, Canberra.

Bruns, T.D., and Read, D.J. 2000. *In vitro* germination of nonphotosynthetic, mycoheterotrophic plants stimulated by fungi isolated from the adult plants. New Phytol. **148:** 335–342.

Bücking, H., and Heyser, W. 1999. Elemental composition and function of polyphosphates in ectomycorrhizal fungi – an X-ray microanalytical study. Mycol. Res. **103:** 31–39.

Bücking, H., and Heyser, W. 2001. Microradiographic localization of phosphate and carbohydrates in mycorrhizal roots of *Populus tremula* × *Populus alba* and the implications for transfer processes in ectomycorrhizal associations. Tree Physiol. **21:** 101–107.

Burgeff, H. 1961. Mikrobiologie des Hochmores. Gustav Fischer Verlag, Stuttgart.

Cairney, J.W.G. 1992. Translocation of solutes in ectomycorrhizal and saprotrophic rhizomorphs. Mycol. Res. **96:** 135–141.

Cairney, J.W.G., and Ashford, A.E. 2002. Biology of mycorrhizal associations of epacrids (Ericaceae). New Phytol. **154:** 305–326.

Cairney, J.W.G., Jennings, D.H., and Agerer, R. 1991.The nomenclature of fungal multi-hyphal linear aggregates. Cryptogam. Bot. **2/3:** 246–251.

Cairney, J.W.G., Rees, B.J., Allaway, W.G., and Ashford, A.E. 1994. A basidiomycete isolated from a *Pisonia* mycorrhiza forms sheathing mycorrhizas with transfer cells on *Pisonia grandis* R. Br. New Phytol. **126:** 91–98.

Caldwell, B.A., Jumpponen, A., and Trappe, J.M. 2000. Utilization of major detrital substrates by dark-septate, root endophytes. Mycologia, **92:** 230–232.

Chilvers, G.A. 1968. Some distinctive types of eucalypt mycorrhiza. Aust. J. Bot. **26:** 49–70.

Cole, L., Orlovich, D.A., and Ashford, A.E. 1998. Structure, function, and motility of vacuoles in filamentous fungi. Fungal Genet. Biol. **24:** 86–100.

Couture, M., Fortin, J.A., and Dalpé, Y. 1983. *Oidiodendron griseum* Robak: an endophyte of ericoid mycorrhiza in *Vaccinium* spp. New Phytol. **95:** 375–380.

Currah, R.S., and Tsuneda, A. 1993. Vegetative and reproductive morphology of *Phialocephala fortinii* (Hyphomycetes, *Mycelium radicis atrovirens*) in culture. Trans. Mycol. Soc. Japan. **34:** 345–356.

Currah, R.S., Zelmer, C.D., Hambleton, S., and Richardson, K.A. 1997. Fungi from orchid mycorrhizas. *In* Orchid Biology: Reviews and perspectives. *Edited by* J. Arditti and A.M. Pridgeon. Kluwer Academic Publishers. Great Britain. pp. 117–170.

Dalpé, Y. 1986. Axenic synthesis of ericoid mycorrhiza in *Vaccinium angustifolium* Ait. by *Oidiodendron* species. New Phytol. **103:** 391–396.

Dalpé, Y. 1989. Ericoid mycorrhizal fungi in the Myxotrichaceae and Gymnoascaceae. New Phytol. **113:** 523–527.

Danielson, R.M., and Pruden, M. 1989. The ectomycorrhizal status of urban spruce. Mycologia, **81:** 335–341.

Declerck, S., Strullu, D.G., and Plenchette, C. 1998. Monoxenic culture of the intraradical forms of *Glomus* sp. isolated from a tropical ecosystem: a proposed methodology for germplasm collection. Mycologia, **90:** 579–585.

Dell, B., Malajczuk, N., and Thomson, G.T. 1990. Ectomycorrhiza formation in *Eucalyptus* V. A tuberculate ectomycorrhiza of *Eucalyptus pilularis*. New Phytol. **114:** 633–640.

Denny, H.J., and Ridge, I. 1995. Fungal slime and its role in the mycorrhizal amelioration of zinc toxicity to higher plants. New Phytol. **130:** 251–257.

Dixon, K. 2003. Underground orchids – on the edge. Plant Talk, **31:** 34–35.

Duc, E., Trouvelot, A., Gianinazzi-Pearson, V., and Gianinazzi, S. 1989. First report of non-nodulating plant mutants (myc-) obtained in pea (*Pisum sativum* L.*)* and Faba bean (*Vicia faba* L.). Plant Sci. **60:** 215–222.

Duckett, J.G., and Read, D.J. 1991. The use of the fluorescent dye 3,3'-dihexyloxacarbocyanine iodide, for selective staining of ascomycete fungi associated with liverwort rhizoids and ericoid mycorrhizal roots. New Phytol. **118:** 259–272.

Duckett, J.G., and Read, D.J. 1995. Ericoid mycorrhizas and rhizoid–ascomycete associations in liverworts share the same mycobiont: isolation of the partners and their resynthesis *in vitro*. New Phytol. **129:** 439–447.

Duckett, J.G., Renzaglia, K.S., and Pell, K. 1991. A light and electron microscope study of rhizoid-ascomycete associations and flagelliform axes in British hepatics. New Phytol. **118:** 233–257.

Duddridge, J.A., Finlay, R.D., Read, D.J., and Söderström, B. 1988. The structure and function of the vegetative mycelium of ectomycorrhizal plants. III. Ultrastructural and autoradiographic analysis of inter-plant carbon distribution through intact mycelia systems. New Phytol. **108:** 183–188.

Duddridge, J.A., and Read, D.J. 1982. An ultrastructural analysis of the development of mycorrhizas in *Monotropa hypopitys* L. New Phytol. **92:** 203–214.

Duddridge, J.A., and Read, D.J. 1984. Modification of the host–fungus interface in mycorrhizas synthesized between *Suillus bovinus* (FR.) O. Kuntz and *Pinus sylvestris* L. New Phytol. **96:** 583–588.

Egger, K.N. 1996. Molecular systematics of E-strain mycorrhizal fungi: *Wilcoxina* and its relationship to *Tricharina* (Pezizales). Can. J. Bot. **74:** 773–779.

Egger, K.N., and Sigler. L. 1993. Relatedness of the ericoid endophytes *Scytalidium vaccinii* and *Hymenoscyphus ericae* inferred from analysis of ribosomal DNA. Mycologia, **85:** 219–230.

Filippi, C., Bagnoli, G., and Giovannetti, M. 1995. Bacteria associated to arbutoid mycorrhizae in *Arbutus unedo* L. Symbiosis, **18:** 57–68.

Finlay, R.D., and Read, D.J. 1986*a*. The structure and function of the vegetative mycelium of ectomycorrhizal plants. I. Translocation of ^{14}C-labelled carbon between ectomycorrhizal plants interconnected by a common mycelium. New Phytol. **103:** 143–156.

Finlay, R.D., and Read, D.J. 1986*b*. The structure and function of the vegetative mycelium of ectomycorrhizal plants. II. The uptake and distribution of phosphorus by mycelial strands interconnecting host plants. New Phytol. **103:** 157–165.

Fortin, J.A., Bécard, G., Declerck, S., Dalpé, Y., St. Arnaud, M., Coughlan, A.P., and Piché, Y. 2002. Arbuscular mycorrhiza on root-organ cultures. Can. J. Bot. **80:** 1–20.

Friese, C.F., and Allen, M.F. 1991. The spread of VA mycorrhizal fungal hyphae in the soil: inoculum types and external hyphal architecture. Mycologia, **83:** 409–418.

Fusconi, A., and Bonfante-Fasolo, P. 1984. Ultrastructural aspects of host–endophyte relationships in *Arbutus unedo* L. mycorrhizas. New Phytol. **96:** 397–410.

Gallaud, I. 1905. Études sur les mycorrhizes endotrophes. Rev. Gen. Bot. **17:** 5–48, 66–83, 123–135, 223–239, 313–325, 425–433, 479–500.

Gao, L.-L., Delp, G., and Smith, S.E. 2001. Colonization patterns in a mycorrhiza-defective mutant tomato vary with different arbuscular-mycorrhizal fungi. New Phytol. **151:** 477–491.

Garbaye, J. 1994. Tansley Review No.76. Helper bacteria: a new dimension to the mycorrhizal symbiosis. New Phytol. **128:** 197–210.

Genre, A., and Bonfante, P. 1997. A mycorrhizal fungus changes microtubule orientation in tobacco root cells. Protoplasma, **199:** 30–38.

Genre, A., and Bonfante, P. 1998. Actin versus tubulin configuration in arbuscule-containing cells from mycorrhizal tobacco roots. New Phytol. **140:** 745–752.

Genre, A., and Bonfante, P. 1999. Cytoskeleton-related proteins in tobacco mycorrhizal cells: γ-tubulin and clathrin localization. Eur. J. Histochem. **43:** 105–111.

Gianinazzi-Pearson, V. 1996. Plant cell responses to arbuscular mycorrhizal fungi: Getting to the roots of the symbiosis. The Plant Cell **8:** 1871–1883.

Giovannetti, M., Sbrana, C., Cisternesi, A.S., and Avio, L. 1996. Analysis of factors involved in fungal recognition responses to host-derived signals by arbuscular mycorrhizal fungi. New Phytol. **133:** 65–71.

Giovannetti, M., Sbrana, C., and Logi, C. 1994. Early processes involved in host recognition by arbuscular mycorrhizal fungi. New Phytol. **127:** 703–709.

Glowa, K.R., Arocena, J.M., and Massicotte, H.B. 2003. Extraction of potassium and and/or magnesium from selected soil minerals by *Piloderma*. Geomicrobiol. J. **20:** 99–111.

Grünig, C.R., Sieber, T.N., Rogers, S.O., and Holdenrieder, O. 2002. Genetic variability among strains of *Phialocephala fortinii* and phylogenetic analysis of the genus *Phialocephala* based on rDNA ITS sequence comparisons. Can. J. Bot. **80:** 1239–1249.

Gunning, B.E.S., and Pate, J. 1974. Transfer cells. *In* Dynamic aspects of plant ultrastructure. *Edited by* A.W. Robards. McGraw Hill, New York. pp. 441–480.

Hagerman, S.M., Sakakibara, S.M., and Durall, D.M. 2001. The potential for woody understory plants to provide refuge for ectomycorrhizal inoculum at an interior Douglas-fir forest after clear-cut logging. Can. J. For. Res. **31:** 711–721.

Hambleton, S., and Currah, R.S. 1997. Fungal endophytes from the roots of alpine and boreal Ericaceae. Can. J. Bot. **75:** 1570–1581.

Hambleton, S., Egger, K.N., and Currah, R.S. 1998. *Oidiodendron*: species delimitation and phylogenetic relationships based on nuclear ribosomal DNA analysis. Mycologia, **90:** 854–869.

Harley, J.L., and Harley, E.L. 1987. A check-list of mycorrhiza in the British Flora. New Phytol. **105:** 1–102.

Harrison, M.J. 1999. Biotrophic interfaces and nutrient transport in plant/fungal symbioses. J. Exp. Bot. **50:** 1013–1022.

Haselwandter, K., and Read, D.J. 1980. Fungal associations of roots of dominant and sub-dominant plants in high-alpine vegetation systems with special reference to mycorrhiza. Oecologia, **45:** 57–62.

Haselwandter, K., and Read, D.J. 1982. The significance of a root–fungus association in two *Carex* species of high-alpine plant communities. Oecologia, **53:** 352–354.

Haug, I., Weber, R., Oberwinkler, F., and Tschen, J. 1991. Tuberculate mycorrhizas of *Castanopsis borneensis* King and *Engelhardtia roxburghiana* Wall. New Phytol. **117:** 25–35.

Heywood, V.H. 1993. Flowering plants of the world. Updated edition, Oxford University Press, 335 pp.

Hirsch, A.M., and Kapulnik, Y. 1998. Signal transduction pathways in mycorrhizal associations: comparisons with the *Rhizobium*-legume symbiosis. Fungal Genet. Biol. **23:** 205–212.

Horton, T.R., Bruns, T.D., and Parker, V.T. 1999. Ectomycorrhizal fungi associated with *Arctostaphylos* contribute to *Pseudotsuga menziesii* establishment. Can. J. Bot. **77:** 93–102.

Ingleby, K., Mason, P.A., Last, F.T., and Fleming, L.V. 1990. Identification of ectomycorrhizas. Institute of Terrestrial Ecology Research Publication No. 5. HMSO, London.

Jabaji-Hare, S., Deschene, A., and Kendrick, B. 1984. Lipid content and composition of vesicles of a vesicular–arbuscular mycorrhizal fungus. Mycologia, **76**: 1024–1030.

Jongmans, A., Van Breemen, N., Lundstrüm, U., Van Hees, P.A.W., Finlay, R.D., Srinivasan, M., Unestam, T., Giesler, R., Melkerud, P.A., and Olsson, M. 1997. Rock-eating fungi. Nature, **389**: 682–683.

Judd, W.S., Campbell, C.S., Kellogg, E.A., and Stevens, P.F. 1999. Plant Systematics, A Phylogenetic Approach. Sinauer Associates, Inc. Sunderland, USA.

Jumpponen, A. 2001. Dark septate endophytes – are they mycorrhizal? Mycorrhiza, **11:** 207–211.

Jumpponen, A., Mattson, K.G., and Trappe, J.M. 1998. Mycorrhizal functioning of *Phialocephala fortinii* with *Pinus contorta* on glacier forefront soil: interactions with soil nitrogen and organic matter. Mycorrhiza, **7:** 261–265.

Jumpponen, A., and Trappe, J.M. 1998. Dark septate endophytes: a review of biotrophic root-colonizing fungi. New Phytol. **140:** 295–310.

Kernan, M.J., and Finocchio, A.F. 1983. A new Discomycete associated with the roots of *Monotropa uniflora* (Ericaceae). Mycologia, **75:** 916–920.

Koske, R.E., Gemma, J.N., and Englander, L. 1990. Vesicular–arbuscular mycorrhizae in Hawaiian-Ericales. Am. J. Bot. **77:** 64–68.

Koske, R.E., Sutton, J.C., and Sheppard. B.R. 1975. Ecology of *Endogone* in Lake Huron sand dunes. Can J. Bot. **53:** 87–93.

Kubota, M., McGonigle, T.P., and Hyakumachi, M. 2001. *Clethra barbinervis*, a member of the order Ericales, forms arbuscular mycorrhizae. Can. J. Bot. **79:** 300–306.

Lamhamedi, M.S., Godbout, C., and Fortin, J.A. 1994. Dependence of *Laccaria bicolor* basidiome development on current photosynthesis of *Pinus strobus* seedlings. Can. J. For. Res. **24:** 1797–1804.

Leake, J.R. 1994. The biology of myco-heterotrophic ('saprophytic') plants. Tansley Review No. 69. New Phytol. **127:** 171–216.

Leake, J.R., and Read, D.J. 1989. The biology of mycorrhiza in the Ericaceae XIII. Some characteristics of the extracellular proteinase activity of the ericoid endophyte *Hymenoscyphus ericae*. New Phytol. **112:** 69–76.

Leake, J.R., and Read, D.J. 1990. Chitin as a nitrogen source for mycorrhizal fungi. Mycol. Res. **94:** 993–1008.

LePage, B.A., Currah, R.S., Stockey, R.A., and Rothwell, G.W. 1997. Fossil ectomycorrhizae from the Middle Eocene. Am. J. Bot. **84:** 410–412.

Leyval, C., Laheurte, F., Belgy, G., and Berthelin, J. 1990. Weathering of micas in the rhizospheres of maize, pine and beech seedlings influenced by mycorrhizal and bacterial inoculation. Symbiosis, **9**: 105–109.

Lindahl, B., Stenlid, J., Olsson, S., and Finlay, R. 1999. Translocation of ^{32}P between interacting mycelia of a wood-decomposing fungus and ectomycorrhizal fungi in microcosm systems. New Phytol. **144:** 183–193.

Ling-Lee, M., Chilvers, G.A., and Ashford, A.E. 1975. Polyphosphate granules in three different kinds of tree mycorrhiza. New Phytol. **75:** 551–554.

Lutz, R.W., and Sjolund, R.D. 1973. *Monotropa uniflora* : ultrastructural details of its mycorrhizal habit. Am. J. Bot. **60:** 339–345.

MacDonald, R.M., Chandler, M.R., and Mosse, B. 1982. The occurrence of bacterium-like organelles in vesicular–arbuscular mycorrhizal fungi. New Phytol. **90:** 659–663.

Massicotte, H.B., Ackerley, C.A., and Peterson, R.L. 1987. Localization of three sugar residues in the interface of ectomycorrhizae synthesized between *Alnus crispa* and *Alpova diplophloeus* as demonstrated by lectin binding. Can. J. Bot. **65:** 1127–1132.

Massicotte, H.B., Melville, L.H., Li, C.Y., and Peterson, R.L. 1992. Structural aspects of Douglas-fir (*Pseudotsuga menziesii* (Mirb.) Franco) tuberculate ectomycorrhizae. Trees, **6:** 137–146.

Massicotte, H.B., Melville, L.H., Molina, R., and Peterson, R.L. 1993. Structure and histochemistry of mycorrhizae synthesized between *Arbutus menziesii* (Ericaceae) and two basidiomycetes, *Pisolithus tinctorius* (Pisolithaceae) and *Piloderma bicolor* (Corticiaceae). Mycorrhiza, **3:** 1–11.

Massicotte, H.B., Melville, L.H., Peterson, R.L., and Molina, R. 1999. Biology of the ectomycorrhizal fungal genus, *Rhizopogon*. IV. Comparative morphology and anatomy of ectomycorrhizas synthesized between several *Rhizopogon* species on Ponderosa pine (*Pinus ponderosa)*. New Phytol. **142:** 355–370.

Massicotte, H.B., Melville, L.H., Peterson, R.L., and Molina, R. 2000. Comparative anatomy of ectomycorrhizas synthesized on Douglas-fir by *Rhizopogon* spp., and the hypogeous relative *Truncocolumella citrina*. New Phytol. **147:** 389–400.

Massicotte, H.B., Melville, L.H., and Peterson, R.L. 2004. Structural features of mycorrhizal associations in two members of the Monotropoideae, *Monotropa uniflora* and *Pterospora andromedea*. Mycorrhiza, **14**. (in press).

Massicotte, H.B., Peterson, R.L., Ackerley, C.A., and Piché, Y. 1986. Structure and ontogeny of *Alnus crispa* – *Alpova diplophloeus* ectomycorrhizae. Can. J. Bot. **64:** 177–192.

Matsubara, Y., Uetake, Y., and R.L. Peterson. 1999. Entry and colonization of *Asparagus officinalis* by arbuscular mycorrhizal fungi with emphasis on changes in host microtubules. Can. J. Bot. **77:** 1159–1167.

McKendrick, S.L., Leake, J.R., Taylor, D.L., and Read, D.J. 2000*a*. Symbiotic germination and development of myco-heterotrophic plants in nature: ontogeny of *Corallorhiza trifida* and characterization of its mycorrhizal fungi. New Phytol. **145:** 523–537.

McKendrick, S.L., Leake, J.R., and Read, D.J. 2000*b*. Symbiotic germination and development of myco-heterotrophic plants in nature: transfer of carbon from ectomycorrhizal *Salix repens* and *Betula pendula* to the orchid *Corallorhiza trifida* through shared hyphal connections. New Phytol. **145:** 539–548.

McLean, C.B., Cunnington, J.H., and Lawrie, A.C. 1999. Molecular diversity within and between ericoid endophytes from the Ericaceae and Epacridaceae. New Phytol. **144:** 351–358.

Melville, L.H., Massicotte, H.B., and Peterson, R.L. 1987. Ontogeny of early stages of ectomycorrhizae synthesized between *Dryas integrifolia* and *Hebeloma cylindrosporum*. Bot. Gaz. **148:** 332–341.

Miller, M.H., McGonigle, T.P., and Addy, H.D. 1995. Functional ecology of vesicular arbuscular mycorrhizas as influenced by phosphate fertilization and tillage in an agricultural ecosystem. Critical Rev. Biotechnol. **15:** 241–255.

Minerdi, D., Fani, R., Gallo, R., and Bonfante, P. 1998. Identification of nitrogen fixation genes in *Burkholderia* endosymbionts of arbuscular mycorrhizal fungi. *In*

Abstracts of the Second International Conference on Mycorrhiza. Uppsala, Sweden. pp. 120–121.

Molina, R., Massicotte, H.B., and Trappe J.M. 1992. Specificity phenomena in mycorrhizal symbioses: community-ecological consequences and practical implications. *In* Mycorrhizal Functioning. *Edited by* M.F. Allen. Chapman & Hall. New York. pp. 357–423.

Molina, R., and Trappe, J.M. 1982. Lack of mycorrhizal specificity by the ericaceous hosts *Arbutus menziesii* and *Arctostaphylos uva-ursi*. New Phytol. **90:** 495–509.

Monreal, M., Berch, S.M., and Berbee, M. 1999. Molecular diversity of ericoid mycorrhizal fungi. Can. J. Bot. **77:** 1580–1594.

Moore, A.E.P., Massicotte, H.B., and Peterson, R.L. 1989. Ectomycorrhiza formation between *Eucalyptus pilularis* Sm., and *Hydnangium carneum*. Wallr. in Dietr. New Phytol. **112:** 193–204.

Morton, J.B., and Redecker, D. 2001. Two new families of *Glomales, Archaeosporaceae* and *Paraglomaceae* with two new genera *Archaeospora* and *Paraglomus*, based on concordant molecular and morphological characters. Mycologia, **93:** 181–195.

Mosse, B. 1970. Honey-coloured, sessile *Endogone* spores: II. Changes in fine structure during spore development. Arch. Mikrobiol. **74:** 129–145.

Mugnier, J., and Mosse, B. 1987. Vesicular–arbuscular mycorrhizal infection in transformed root-inducing T-DNA roots grown axenically. Phytopathology, **77:** 1045–1050.

Narisawa, K., Kawamata, H., Currah, R.S., and Hashiba, T. 2002. Suppression of *Verticillium* wilt in eggplant by some fungal root endophytes. Eur. J. Plant Pathol. **108:** 103–109.

Newsham, K.K. 1999. *Phialophora graminicola*, a dark septate fungus, is a beneficial associate of the grass *Vulpia ciliata* ssp. *ambigua*. New Phytol. **144:** 517–524.

Nurmiaho–Lassila, E.-L., Timonen, S., Haahtela, K., and Sen, R. 1997. Bacterial colonization patterns of intact *Pinus sylvestris* mycorrhizospheres in dry pine forest soil: an electron microscopy study. Can. J. Microbiol. **43:** 1017–1035.

O'Dell, T.E., Massicotte, H.B., and Trappe, J.M. 1993. Root colonization of *Lupinus latifolius* Agardh., and *Pinus contorta* Dougl. by *Phialocephala fortinii* Wang & Wilcox. New. Phytol. **124:** 93–100.

Olsson, P.A., and Wallander, H. 1998. Interactions between ectomycorrhizal fungi and the bacterial community in soils amended with various primary minerals. FEMS Microbiol. Ecol. **27:** 195–205.

Orlovich, D.A., and Ashford, A.E. 1993. Polyphosphate granules are an artefact of specimen preparation in the ectomycorrhizal fungus *Pisolithus tinctorius*. Protoplasma, **173:** 91–102.

Paris, F., Bonnaud, P., Ranger, J., and Lapeyrie, F. 1995. *In vitro* weathering of phlogopite by ectomycorrhizal fungi. I. Effect of K^+ and Mg^{2+} deficiency on phyllosilicate evolution. Plant and Soil, **177:** 191–201.

Peretto, R., Perotto, S., Faccio, A., and Bonfante-Fasolo, P. 1990. Cell Surface in *Calluna vulgaris* L. hair roots. *In situ* localization of polysaccharidic components. Protoplasma, **155:** 1-18.

Perotto, S., Actis-Perino, E., Perugini, J., and Bonfante, P. 1996. Molecular diversity of fungi from ericoid mycorrhizal roots. Mol. Ecol. **5:** 123–131.

Perotto, S., Girlanda, M., and Martino, E. 2002. Ericoid mycorrhizal fungi: some new perspectives on old acquaintances. Plant Soil, **244:** 41–53.

Perotto, S., Peretto, R., Faccio, A., Schubert, A., Varma, A., and Bonfante, P. 1995. Ericoid mycorrhizal fungi: cellular and molecular bases of their interactions with the host plant. Can. J. Bot. **73 (Suppl.):** S557–S568.

Peterson, R.L., and Bradbury, S.M. 1995. Use of plant mutants, intraspecific variants, and non-hosts in studying mycorrhiza formation and function. *In* Mycorrhiza Structure, function, molecular biology and biotechnology. *Edited by* A. Varma and B. Hock. Springer-Verlag, Berlin. pp. 157–180.

Peterson, R.L., and Guinel, F.C. 2000. The use of plant mutants to study regulation of colonization by AM fungi. *In* Arbuscular mycorrhizas: physiology and function. *Edited by* Y. Kapulnik and D.D. Douds, Jr. Kluwer Academic Publishers, Dodrecht. pp. 147–171.

Peterson, R.L., Bonfante, P., Faccio, A., and Uetake, Y. 1996. The interface between fungal hyphae and orchid protocorm cells. Can. J. Bot. **74:** 1861–1870.

Peterson, R.L., Uetake, Y., and Zelmer, C. 1998. Fungal symbioses with orchid protocorms. Symbiosis, **25:** 29–55.

Randall, B.L., and Grand, L.F. 1986. Morphology and possible mycobiont (*Suillus pictus*) of a tuberculate ectomycorrhiza on *Pinus strobus*. Can. J. Bot. **64:** 2182–2191.

Rasmussen, H.N. 1995. Terrestrial orchids from seed to mycotrophic plant. Cambridge University Press, Cambridge.

Read, D.J. 1974. *Pezizella ericae* sp. nov., the perfect state of a typical mycorrhizal endophyte of Ericaceae. Trans. Br. Mycol. Soc. **63:** 381–419.

Read, D.J. 1978. Biology of mycorrhiza in heathland ecosystems with special reference to the nitrogen nutrition of Ericaceae. *In* Microbial Ecology. *Edited by* M.W. Loutit and J.A.R. Miles. Springer-Verlag, Berlin. pp. 324–328.

Read, D.J. 1984. The structure and function of the vegetative mycelium of mycorrhizal roots. *In* The ecology and physiology of the fungal mycelium. *Edited by* D.H. Jennings and A.D.M. Rayner. Cambridge University Press, Cambridge. pp. 215–240.

Read, D.J. 1996. The structure and function of the ericoid mycorrhizal root. Ann. Bot. **77:** 365–376.

Read, D.J., and Haselwandter, K. 1981. Observations on the mycorrhizal status of some alpine plant communities. New Phytol. **88:** 341–352.

Read, D.J., Duckett, J.G., Francis, R., Ligrone, R., and Russell, A. 2000. Symbiotic fungal associations in 'lower' land plants. Phil. Trans. R. Soc. Lond. B **355:** 815–831.

Redecker, D., Kodner, R., and Graham, L.E. 2000. Glomalean fungi from the Ordovician. Science, **289:** 1920–1921.

Robertson, D.C., and Robertson, J.A. 1982. Ultrastructure of *Pterospora andromedea* Nuttall and *Sarcodes sanguinea* Torrey mycorrhizas. New Phytol. **92:** 539–551.

Robertson, D.C., and Robertson, J.A. 1985. Ultrastructural aspects of *Pyrola* mycorrhizae. Can. J. Bot. **63:** 1089–1098.

Ruiz-Lozano, J.M., and Bonfante, P. 1999. Identification of a putative P-transporter operon in the genome of a *Burkholderia* strain living inside the arbuscular mycorrhizal fungus *Gigaspora margarita*. J. Bacteriol. **181:** 4106–4109.

Sarand, I., Timonen, S., Nurmiaho-Lassila, E.-L., Koivula, T., Haahtela, K., Romantschuk, M., and Sen, R. 1998. Microbial biofilms and catabolic plasmid harbouring degradative fluorescent pseudomonads in Scots pine mycorrhizospheres developed on petroleum contaminated soils. FEMS Microbiol. Ecol. **27:** 115–126.

Scales, P.F., and Peterson, R.L. 1991*a*. Structure of ectomycorrhizae formed by *Wilcoxina mikolae* var. *mikolae* with *Picea mariana* and *Betula alleghaniensis*. Can. J. Bot. **69** : 2149–2157.

Scales, P.F., and Peterson, R.L. 1991*b*. Structure and development of *Pinus banksiana – Wilcoxina* ectendomycorrhizae. Can. J. Bot. **69:** 2135–2148.

Scannerini, S., and Bonfante, P. 1991. Bacteria and bacteria-like objects in endomycorrhizal fungi (Glomaceae). *In* Symbiosis as a source of evolutionary innovation: speciation and morphogenesis. *Edited by* L. Margulis and R. Fester. MIT Press, Cambridge, Mass. pp. 273–287.

Schadt, C.W., Mullen, R.B., and Schmidt, S.K. 2001. Isolation and phylogenetic identification of a dark-septate fungus associated with the alpine plant *Ranunculus adoneus*. New Phytol. **150:** 747–755.

Schmid, E., and Oberwinkler, F. 1993. Mycorrhiza-like interaction between the achlorophyllous gametophyte of *Lycopodium clavatum* L., and its fungal endophyte studied by light and electron microscopy. New Phytol. **124:** 69–81.

Schmid, E., and Oberwinkler, F. 1995. A light- and electron-microscopic study on a vesicular–arbuscular host–fungus interaction in gametophytes and young sporophytes of the Gleicheniaceae (Filicales). New Phytol. **129:** 317–324.

Schüßler, A., Schwarzott, D., and Walker, C. 2001. A new fungal phylum, the *Glomeromycota*: phylogeny and evolution. Mycol. Res. **105:** 1413–1421.

Senoo, K., Solaiman, M.Z., Kawaguchi, M., Imaizumi-Anraku, H., Akao, S., Tanaka, A., and Obata, H. 2000. Isolation of two different phenotypes of mycorrhizal mutants in the model legume plant *Lotus japonicus* after EMS-treatment. Plant Cell Physiol. **41:** 726–732.

Shepherd, V.A., Orlovich, D.A., and Ashford, A.E. 1993*a*. Cell-to-cell transport via motile tubules in growing hyphae of a fungus. J. Cell. Sci. **105:** 1171–1178.

Shepherd, V.A., Orlovich, D.A., and Ashford, A.E. 1993*b*. A dynamic continuum of pleiomorphic tubules and vacuoles in growing hyphae of a fungus. J. Cell. Sci. **104:** 495–507.

Smith, F.A., and Smith, S.E. 1997. Tansley Review No. 96. Structural diversity in (vesicular)–arbuscular mycorrhizal symbioses. New Phytol. **137:** 373–388.

Smith, S.E., and Read, D.J. 1997. Mycorrhizal Symbiosis. Academic Press, London.

Smith, S.E., and Smith, F.A. 1990. Structure and function of the interfaces in biotrophic symbioses as they relate to nutrient transport. New Phytol. **114:** 1–38.

St. Arnaud, M., Hamel, C., Vimard, B., Caron, M., and Fortin, J.A. 1996. Enhanced hyphal growth and spore production of the arbuscular mycorrhizal fungus *Glomus intraradices* in an *in vitro* system in absence of host roots. Mycol. Res. **100:** 328–332.

Stoyke, G., Egger, K.N., and Currah, R.S. 1992. Characterization of sterile endophytic fungi from the mycorrhizae of subalpine plants. Can. J. Bot. **70:** 2009–2016.

Strullu, D.G., Harley, J.L., Gourret, J.P., and Garrec, J.P. 1982. Ultra-structure and microanalysis of the polyphosphate granules of the ectomycorrhizas of *Fagus sylvatica*. New Phytol. **92:** 417–423.

Taylor, D.L., and Bruns, T.D. 1997. Independent, specialized invasions of ectomycorrhizal mutualism by two nonphotosynthetic orchids. Proc. Natl. Acad. Sci. USA **94:** 4510–4515.

Taylor, D.L., and Bruns, T.D. 1999. Population, habitat and genetic correlates of mycorrhizal specialization in the 'cheating' orchids *Corallorhiza maculata* and *C. mertensiana*. Mol. Ecol. **8:** 1719–1732.

Taylor, T.N., Remy, W., Hass, H., and Kerp, H. 1995. Fossil arbuscular mycorrhizae from the Early Devonian. Mycologia, **87:** 560–573.

Thiers, H.D. 1975. California mushrooms, a field guide to the Boletes. Hafner Press, New York.

Timonen, S., and Peterson, R.L. 2002. Cytoskeleton in mycorrhizal symbiosis. Plant Soil **244:** 199–210.

Uetake, Y., Farquhar, M.L., and Peterson, R.L. 1997. Changes in microtubule arrays in symbiotic orchid protocorms during fungal colonization and senescence. New Phytol. **135:** 701–709.

Uetake, Y., and Ishizaka, N. 1996. Cytochemical localization of adenylate cyclase activity during the symbiotic development of *Spiranthes sinensis* induced by the symbiotic fungus, *Ceratobasidium cornigerum*. Mycol. Res. **96:** 105–112.

Uetake, Y., and Peterson, R.L. 1997. Changes in microfilament arrays in protocorm cells of the orchid species, *Spiranthes sinensis,* induced by the symbiotic fungus, *Ceratobasidium cornigerum*. Can. J. Bot. **75:** 1661–1669.

Uetake, Y., and Peterson, R.L. 1998. Association between microtubules and symbiotic fungal hyphae in protocorm cells of the orchid species, *Spiranthes sinensis*. New Phytol. **140:** 715–722.

Uetake, Y., and Peterson, R.L. 2000. Spatial associations between actin filaments, endoplasmic reticula, mitochondria and fungal hyphae in symbiotic cells of orchid protocorms. Mycoscience, **41:** 481–489.

Ursic, M., Peterson, R.L., and Husband, B. 1997. Relative abundance of mycorrhizal fungi on *Pinus strobus* seedlings affected by root rot in a southern Ontario nursery. Can. J. For. Res. **27:** 54–62.

Vesk, P.A., Ashford, A.E., Markovina, A-L., and Allaway, W.G. 2000. Apoplasmic barriers and their significance in the exodermis and sheath of *Eucalyptus pilularis – Pisolithus tinctorius* ectomycorrhizas. New Phytol. **145:** 333–346.

Villegas, J., Williams, R.D., Nantais, L., Archambault, J., and Fortin, J.A. 1996. Effects of N source on pH and nutrient exchange of extramatrical mycelium in a mycorrhizal Ri T-DNA transformed root system. Mycorrhiza, **6:** 247–251.

Vreeland, P., Kleiner, E.F., and Vreeland, H. 1981. Mycorrhizal symbiosis of *Sarcodes sanguinea*. Environ. Exptl. Bot. **21:** 15–25.

Wallander, H., and Wickman, T. 1999. Biotite and microcline as potassium source in ectomycorrhizal and non-ectomycorrhizal *Pinus sylvestris* seedlings. Mycorrhiza, **9:** 25–32.

Wallander, H.,Wickman, T., and Jacks, G. 1997. Apatite as a P source in mycorrhizal and non-mycorrhizal *Pinus sylvestris* seedlings. Plant Soil, **196:** 123–131.

Warcup, J.H. 1985. *Rhizanthella gardneri* (Orchidaceae), its *Rhizoctonia* endophyte and close association with *Melaleuca uncinata* (Myrtaceae) in Western Australia. New Phytol. **99:** 273–280.

Williamson, B., and Hadley, G. 1969. DNA content of nuclei in orchid protocorms infected with *Rhizoctonia*. Nature, **222:** 582–583.

Xiao, G., and Berch, S.M. 1995. The ability of known ericoid mycorrhizal fungi to form mycorrhizae with *Gaultheria shallon*. Mycologia, **87:** 467–470.

Xiao, G., and Berch, S.M. 1996. Diversity and abundance of ericoid mycorrhizal fungi of *Gaultheria shallon* on forest clearcuts. Can. J. Bot. **74:** 337–346.

Young, B.W., Massicotte, H.B., Tackaberry, L.E., Baldwin, Q.F., and Egger, K.N. 2002. *Monotropa uniflora*: morphological and molecular assessment of mycorrhizae retrieved from sites in the sub-boreal spruce biogeoclimatic zone in central British Columbia. Mycorrhiza, **12:** 75–82.

Young, J.A., and Young, C.G. 1992. Seeds of woody plants in North America, revised and enlarged edition. Dioscorides Press, 407 pp.

Yu, T. E. J-C., Egger, K.N., and Peterson R.L. 2001*a*. Ectendomycorrhizal associations – characteristics and functions. Mycorrhiza, **11:** 167–177.

Yu, T., Nassuth, A., and Peterson, R.L. 2001*b*. Characterization of the interaction between the dark septate fungus *Phialocephala fortinii* and *Asparagus officinalis* roots. Can. J. Microbiol. **47:** 741–753.

Zak, B. 1976. Pure culture synthesis of bearberry mycorrhizae. Can. J. Bot. **54:** 1297–1305.

Zelmer, C.D. 2001. Root-associated organisms of the Cypripedioideae (Orchidaceae). Ph.D. thesis. University of Guelph, Guelph, Ont. 227 pp.

Zelmer, C.D., and Currah, R.S. 1995. Evidence for a fungal liaison between *Corallorhiza trifida* (Orchidaceae) and *Pinus contorta* (Pinaceae). Can. J. Bot. **73:** 862–866.

Zhi-wei, Z. 2000. The arbuscular mycorrhizas of pteridophytes in Yunnan, southwest China: evolutionary interpretations. Mycorrhiza, **10:** 145–149.

Glossary

apoplast: The non-living portion of fungal and plant cells.

appressorium: A swelling on a germ tube or hypha involved in attachment of hypha to root surface.

arbusculate coil: Coils of fungal hyphae with fine lateral branches; typical of *Paris*-type arbuscular mycorrhizas.

arbuscule: A finely branched hypha forming a tree-like structure within a root cortical cell.

autotrophic: Able to manufacture organic compounds from inorganic compounds in the environment.

auxiliary cells (vesicles; bodies): Spore-like structures formed in the extraradical mycelium in species of the arbuscular mycorrhiza genera *Gigaspora* and *Scutellospora*.

basidiocarp: A fungal reproductive structure found in the Basidiomycota involved in producing basidiospores.

biofilm: Microorganisms (generally bacteria) associated with the surface of roots or fungal hyphae.

callose: A complex branched polysaccharide often found at sites of pathogen attack.

Casparian band: A region of primary wall containing suberin and lignin found in the root endodermis and exodermis making the wall impervious to water; named after Robert Caspary.

chlamydospore: An asexual spore.

cytoskeleton: Subcellular network consisting of microtubules and actin filaments.

cystidia: Sterile hyphae, often with distinctive shapes, emanating from the suface of the mantle of some ectomycorrhizas, arbutoid mycorrhizas and monotropoid mycorrhizas.

dematiaceous hyphae: Pigmented hyphae, usually with melanin in the cell wall.

depletion zone: The region around a root in which nutrients such as phosphorus have been absorbed.

endodermis: The inner layer of cortical cells in the root that develops a Casparian band.

endophyte: Organism that colonizes plant tissues without causing any apparent injury.

entry point: In arbuscular mycorrhizas, a site along the root axis at which an hypha enters.

epigeous: Occurring on the surface of the ground.

exodermis: The outer layer of cortical cells in the root that develop Casparian bands and usually suberin lamellae.

extraradical mycelium: Fungal hyphae that emanate from the surface of a mycorrhiza and grow into the substrate.

fruit-bodies: A general term for reproductive bodies of many fungi.

fungal peg: A specialized hypha surrounded by finger-like projections of host cell wall within an epidermal cell in monotropoid mycorrhizas, presumably for bi-directional transfer of nutrients.

germ tube: A hypha growing from a fungal spore.

glomalin: A glycoprotein produced by arbuscular mycorrhiza hyphae that has a role in soil aggregation.

glycogen: A polysaccharide stored in fungal cells.

hair roots: Very fine roots with simple anatomy found in the family Ericaceae that become colonized by ericoid fungi.

Hartig net: Fungal hyphae that grow between epidermal and/or cortical cells.

heterotrophic: Unable to manufacture organic compounds from inorganic compounds.

hyaline: Lacking pigment.

hyphal coil: A hypha that forms a loop within a root cell.

hyphal complex: A coiled, highly branched hypha found within host cells.

hypogeous: Occurring under the surface of the ground.

interfacial matrix: Host-derived cell wall constituents between the perifungal membrane and hyphal cell wall.

labyrinthine: The complex branching pattern shown by hyphae of the inner mantle and Hartig net.

mantle: A sheath of fungal hyphae enclosing a root; found in ectomycorrhizas, ectendomycorrhizas, arbutoid mycorrhizas and monotropoid mycorrhizas.

microsclerotium: An aggregation of hyphae containing storage materials within epidermal or cortical cells.

middle lamella: A pectin rich layer between the primary walls of adjacent cells.

morphotype: A morphologically distinctive type of ectomycorrhiza, ectendomycorrhiza, arbutoid or monotropoid mycorrhiza.

myco-heterotrophic plants: Plants that depend on links with fungi for their source of carbon compounds.

peloton: Specialized hyphal complex in orchid host cells.

periarbuscular membrane: Host-derived plasma membrane that surrounds branches of an arbuscule.

perifungal membrane: Host-derived plasma membrane that surrounds fungal hyphae within a plant cell.

plasmodesmata: Cytoplasmic strands that connect the protoplasts of living plant cells.

polyphosphate: A storage form of phosphate found in fungal vacuoles.

protocorm: In orchids, the stage between embryo and seedling.

rhizoid: An epidermal outgrowth in lower plants involved in nutrient acquisition.

rhizomorph: A linear aggregation of hyphae emanating from the surface of the mantle of some mycorrhizas.

rhizosphere: The substrate immediately surrounding roots.

runner hyphae: Fungal hyphae in arbuscular mycorrhizas that grow rapidly along the surface of a root.

saprotroph (saprobe): An organism that gains nutrients from non-living organic matter.

sclerotium: A mass of hyphae originating in the extraradical mycelium of some mycorrhizas that is organized into a rounded structure covered by melanized hyphae (the rind) enclosing hyphae with various storage compounds.

suberin lamellae: Layers of suberin, a fatty material, deposited against the primary cell wall in the endodermis and exodermis of some roots.

suspensor: A structure at the base of the embryo in most vascular plants.

symplast: The living portion of fungal and plant cells.

tilosome: A complex of wall projections in velamen cells adjacent to short cells of the exodermis of orchid roots.

transfer cell: Parenchyma cell that becomes modified by the formation of ingrowths of primary cell wall enveloped by plasma membrane; found in root cells adjacent to Hartig net hyphae in some ectomycorrhizas.

transformed root: A root that has had the *Agrobacterium tumefaciens* or *Agrobacterium rhizogenes* genome inserted into the nuclei of its cells.

tubercle: Cluster of ectomycorrhizas enclosed in a sheath (peridium) of hyphae.

velamen: A multi-layered epidermis in aerial and some terrestrial roots of orchids.

vesicle: A lipid-filled enlarged portion of hypha that may develop a thickened wall.

vessel hyphae: Hyphae of large diameter located within rhizomorphs involved with transport of water and mineral ions.

Appendices

The methods included as appendices are brief outlines of those used in the preparation of the images in the preceding chapters. For detailed methods of preparing roots and mycorrhizas for microscopic examination, several sources should be consulted:

1. Brundrett, M., Bougher, N., Dell, B., Grove, T., and Malajczuk, N. 1996. Working With Mycorrhizas in Forestry and Agriculture. ACIAR Monograph 32.
2. Brundrett, M., Melville, L., and Peterson, L. 1994. Practical Methods in Mycorrhiza Research. Mycologue Publications Ltd., Waterloo.
3. Norris, J.R., Read, D.J., and Varma, A.K. (*Editors*). 1991. Methods in Microbiology. Vol. 23. Academic Press Ltd., London.
4. Ruzin, S.E. 1999. Plant Microtechnique and Microscopy. Oxford University Press. New York, Oxford.

Appendix A: Clearing and staining roots for examination of arbuscular mycorrhizas, ericoid mycorrhizas, orchid mycorrhizas, and dark septate endophytes:

1. Roots should be cleaned of soil and are usually fixed in 50% ethanol for at least 24 h. Roots can be cleared without prior fixation.
2. Rinse 3× in water (deionized if available).
3. Clear in either 5% or 10% KOH (depending on the fragility of roots) by either autoclaving for 20–40 min or placing containers in a water bath at 90°C for 2–3 h. Some very fine roots can be cleared at room temperature if left overnight or longer.
4. Rinse 3× in water (deionized if available).
5. If roots are very pigmented they will need to be bleached (30–35% hydrogen peroxide:distilled water (1:1) and then add ammonium hydroxide for a final concentration of 0.05%). Length of time will depend on roots but usually about 5 min is sufficient.
6. Rinse 3× in water.
7. Acidify in 2% HCl or 2% lactic acid for 1–2 min.
8. Stain in **either** 0.03% Chlorazol Black E (made up in a solution of 80% lactic acid: glycerin in distilled water 1:1:1) for 2–3 h at 90°C **or** 0.05% Trypan blue made up as for Chlorazol Black E **or** 0.1% acid fuchsin made up in the same way. Staining with the latter two stains may require more time. Staining with acid fuchsin has been particularly useful when material is to be examined by confocal laser scanning microscopy.
9. De-stain in 50% glycerin for approx. 24 h.
10. Mount roots on slides in 50% glycerin.

Appendix B: Embedding material for light microscopy:

To obtain uniform thin sections of a specimen it is necessary to embed the material in resin so that it can be sectioned with glass or diamond knives using a microtome. After infiltration and polymerization in a resin, the tissue and the surrounding medium is of equal density. There are many protocols for this type of work, depending on the type of investigation (see Ruzin 1999). For most light microscopic work, we have routinely used the following method:

1. Prepare 2.5% glutaraldehyde in either 0.1M HEPES or 0.1M Sorensen's phosphate buffer and adjust pH to 6.8. This must be done in a FUME HOOD.
2. Cut specimens into small pieces (usually a few mm in length) and place in vials containing buffered glutaraldehyde, again in a FUME HOOD. Fix tissues for a minimum of 3 h at room temperature (can be left at 4°C overnight or longer).
3. Remove the glutaraldehyde with a pasteur pipette in a FUME HOOD, placing this solution in a waste-container.
4. Dehydrate the tissue by means of a graded ethanol series:

 50% for 20 min
 70% for 20 min
 90% for 30 min
 95% for 30 min
 100% for 30 min
 100% for 45 min
 100% for 60 min (store overnight in 100% ethanol if necessary)

5. Gradually infiltrate with LR White resin (Obtained from London Resin Company, P.O. Box 34, Basingstoke, Hampshire RG21 2NW, United Kingdom, or from MARIVAC LTD., Halifax, N.S., Canada). We use the Medium grade.
6. Ratio of 100% ethanol to LR White resin:

 2 : 1 for 30 min
 1 : 1 for 30 min
 1 : 2 for 30 min
 100% LR White for 60 min
 100% LR White for 60 min
 100% LR White overnight.

7. Embedding in gelatin capsules:
 This is a common method for LR White resin, as LR White polymerizes with heat in the absence of air. Capsules of 6.5 mm diameter (#1) are convenient. We usually make a holder for the capsules out of small, light cardboard boxes with holes the diameter of the capsule poked in the top side with a pencil or similar instrument. Remove the tops from the capsules and place the bottom half in the holes. Place fresh LW White resin in the capsules.
 Transfer one specimen into each capsule with a Pasteur pipette and then top off the capsule with fresh resin. It may be easier to pour all the contents (e.g. root tips) of the vial into a small disposable dish and carefully transfer each specimen with a toothpick.
 Replace the capsule lid tightly, remember that LR White polymerizes in the absence of oxygen, so that as little air as possible remains in the top of the capsule. When the samples are deposited in the capsules, indicate on the box which specimens you have, and then place them in the oven at 60°C.

Samples usually polymerize overnight. This method works consistently, but the samples may end up in a twisted configuration at the bottom of the capsule. These need to be cut out and re-mounted on a new resin block in the desired orientation.

8. Flat embedding with LR White:
   When it is necessary to embed specimens on a flat surface (as in the case of long thin roots found in most AMs) then the gelatin capsule method is unreliable. In this case transfer the resin infiltrated samples into an aluminum weighing dish and cover them with at least 2 mm of fresh resin. Place another aluminum weighing dish on top of the resin so that a seal is formed. Mark the specimens with a small strip of paper with a code written in pencil placed in the resin. Separate the specimens in the weighing dish with a toothpick so that they don't clump together during polymerization. Place dishes in a 60°C oven and polymerize for 3–12 h.
   Make sure the weighing dish or embedding molds are lying flat in the oven, or the samples will clump together at the lowest point.
9. Flat embedding with LR White polymerized with UV light:
   LR White can be polymerized with UV light; the protocol for this is the same as in **8** but the weighing dishes are placed under a UV light source.
10. When using the flat embedding methods, specimens are cut out of the resin using a small saw and mounted on resin stubs to fit the chuck of the microtome.

Appendix C: Embedding tissue for transmission electron microscopy:

There are many methods of fixing and embedding tissues for transmission electron microscopy. We routinely use the following protocol:

1. Cut tissues into small pieces and fix with 2.5% glutaraldehyde in either buffer listed under Appendix B. This should be done in small vials on ice in a FUME HOOD. Fix for 1 to several h depending on the size of the sample.
2. Rinse with ice cold buffer 3× over a 1 h period.
3. Working in a fume hood, post-fix sample in 2% osmium tetroxide either in buffer or in water for 2 h to overnight at 4°C. Use only enough osmium to cover sample. Prepare 2% osmium tetroxide in advance by adding one 0.5 g vial of osmium tetroxide to 25 mL distilled water or buffer.
   (Use extreme caution, osmium will fix tissues instantly, especially your eyes. Work in the fume hood and wear gloves)
4. Remove the osmium and place it in a waste container. Rinse tissue in buffer several times over 1 h on ice.
5. Dehydrate sample in a graded acetone or ethanol series for a minimum of 30 min for each step or preferably longer (20%, 40%, 60%, 80%, 95%, 100%).
6. Make 3 changes of 100% acetone or ethanol then gradually add Spurr's resin over 1–2 days.
   Spurr's resin should be made just prior to use as follows:
   Place a plastic disposable beaker on a top-loading scale and add in order:
   vinylcyclohexene dioxide (VCD) 10.0 g
   diglycidyl ether of propylene glycol 6.0 g
   nonenyl succinic anhydride (NSA) 26.0 g
   (*mix the above ingredients thoroughly before adding the last ingredient*)
   dimethyl aminoethanol (DMAE) 0.2 g

7. Evaporate acetone or ethanol off overnight by covering vials with perforated aluminum foil.
8. Embed in fresh resin in aluminum weighing boats and polymerize at 60°C for 24–36 h.

Appendix D: Preparation of samples for scanning electron microscopy

1. Fix tissue in 2.5% glutaraldehyde in 0.07M phosphate buffer at pH 6.8 for 3–24 h, depending on the size of the sample. Fixation can be enhanced if carried out on a rotary shaker. Work in the fume hood and wear latex gloves.
2. Remove waste glutaraldehyde carefully with Pasteur pipette and dispose of it in a proper waste container.
3. Wash the samples in phosphate buffer for 5 min and repeat. Dispose of waste buffer in glutaraldehyde waste container.
4. Post-fix tissue in 1% osmium tetroxide in phosphate buffer (or water) at room temperature.Use osmium sparingly as it is very expensive.
5. Wash the tissue twice in phosphate buffer then rinse the tissue 5× in distilled water (d H_2O). Steps 6–10 are optional.
6. Prepare a solution of 1% thiocarbohydrazide in d H_2O immediately before using. Dissolve the thiocarbohydrazide crystals in d H_2O by stirring for at least 10 min. Filter the solution using a filter paper and funnel.
7. Place the washed samples in the thiocarbohydrazide solution for 30 min.
8. Wash the tissue several times in d H_2O.
9. Post-fix the tissue in 2% osmium tetroxide in buffer or d H_2O for 1 h.
10. Remove osmium and rinse 2× in d H_2O.
11. Dehydrate samples with a graded series of ethanol (30%, 50%, 70%, 90%, 95%, 100%) for 30 min each.
12. Samples are now ready for critical point drying.

Appendix E: Staining resin-embedded tissues for light microscopy

Samples embedded in either LR White or Spurr's resin can be stained for light microscopy with a number of methods (see Ruzin 1999). We routinely use the following:

1. Prepare a solution of Toluidine Blue O (TBO) by adding 0.05 g of TBO and 1.0 g sodium borate to 100 mL d H_2O. Filter and store in a reagent bottle.
2. After sections have been heat-fixed to a slide, add TBO to cover sections and heat gently.
3. Rinse stain from slide and let dry.
4. Permanent mounts can be prepared by adding a drop of immersion oil and a cover glass and then sealing the edges of the cover glass with nail polish.

Appendix F: Staining hand sections for light microscopy

A considerable amount of information can be obtained concerning root and mycorrhiza structure using fresh material. A detailed description of how to prepare hand sections and some of the stains that are useful can be found in Brundrett et al. (1994, 1996).

Appendix G: Staining tissues for confocal laser scanning microscopy

For studying general morphological characteristics of fungal structures within roots in arbuscular mycorrhizas, ericoid mycorrhizas, orchid mycorrhizas and dark septate fungal endophytes, the following procedure works very well. Collect root samples and store at least overnight in 50% ethanol. Clear roots by heating in 5% KOH (w/v) in a water bath at 90°C, rinse roots with water, and place them in 0.1N HCl for 5–10 min. Stain roots in 0.01% acid fuchsin (w/v) in a solution of lactic acid–glycerin–water (875 mL lactic acid, 63 mL glycerin, 63 mL water) for 1 h in a water bath at 55°C. Place roots in 100% glycerin for destaining. Best results are obtained with confocal microscopy when the fungal structures are only very lightly stained.

Index